FITTING OUT
A MOULDED HULL

BY THE SAME AUTHOR

ABC of Gliding
Handbook of Small Boat Cruising

Fitting Out
a Moulded Hull

FOX GEEN

Illustrations
by Maureen Verity

HOLLIS & CARTER
LONDON SYDNEY
TORONTO

As always
TO SLIM

© A. Fox-Geen 1974

This book is sold subject to the condition that it shall not, by way of trade or otherwise, be lent, re-sold, hired out, or otherwise circulated without the publisher's prior consent in any form of binding or cover other than that in which it is published and without a similar condition including this condition being imposed on the subsequent purchaser.

Paperback edition I S B N 0 370 10314 9
Hard-cover edition I S B N 0 370 10267 3
Printed in Great Britain for
Hollis & Carter
an associate company of
The Bodley Head Ltd
9 Bow Street, London WC2E 7AL
by Northumberland Press Limited
Gateshead
Set in Linotype Baskerville
*First published as a paperback and
simultaneously as a hard-cover edition 1974*

CONTENTS

Author's Foreword, 9

1
Resinglass and Boats, 11

Resinglass, 12 · Glass cloth, 13 · Resin, 15 · Polyester mixes, 16 · Types of Polyester resin, 19 · Using resinglass, 20 · Bonding to cured surfaces, 25 · Bonding to overhead surfaces, 26 · Bonding to wood, 27 · Bonding to metal, 27 · Use of cloth for bonding, 28 · Other methods of attachment, 32 · Sandwich construction, 33 · Treatment of cured resinglass, 34 · Cleanliness in working, 36 · Decoration, 39 · Fire risk, 41

2
Deciding on a Hull, 42

Hull moulding, 43 · Design, 44 · Moulding, 45 · Availability, 45 · Handling a new hull, 47 · Purpose, 48 · Seaworthiness, 48 · Handling characteristics, 48 · Performance, 49 · Habitability, 49 · Appearance, 50 · Rig, 51 · Power, 51 · Price, 52 · Value, 53

3
Buying, 54

Emulate the squirrel, 55 · Tools, 57

4
Planning, 60

The working schedule, 61 · Spreading the load, 61 · A working schedule in detail, 62

5
The Working Site, 65

Facilities, 65 · Working base, 66 · Inspection, 69

6
Preparatory Work, 70

Tools, 70 · Marking up the hull, 70 · Transferring the Load Waterline, 72 · Perpendiculars and verticals, 72 · Symmetry, 74 · Location of components, 74 · Exterior markings, 76 · Drainage, 76

7
Sub-Mouldings, 77

Moulding a hatch cover, 77 · Routines, 81 · Laying-up, 83 · Release, 84 · To mould in a water tank, 84

8
Starting to Fit Out, 86

Rubbing strakes, 86 · Handrails and grabs, 90 · Keels, 90 · Cockpit drains, 93 · Hatch covers, 93 · Leaks, 97

9
On Deck, 98

Ventilation, 98 · Ground tackle fittings, 99 · Toe-rails, 99 · Coamings, 100 · Winches, 101 · Chain-plates, 102 · Mast steps, 104 · Through masts, 104 · Tracking, 105 · Pulpits and stanchions, 105 · Bowsprits and bumpkins, 107 · Rudder fittings, 108 · Safety fittings, 110

10
Below Decks, 111

Wiring and piping, 111 · Templates, 113 · Fitting a bulkhead, 116 · Stringers, 118 · Framing, 119 · Galley, 120 · Lockerage, 121 · Cabin sole, 124 · The heads, 125 · Cable stowage, 125 · Ventilation, 127 · Tankage, 128 · Decoration, 129 · Condensation, 129 · Windows and other lights, 131 · Non-slip decks, 133

CONTENTS

11
Engines, 134
Inboards, 134 · Batteries, 135 · Fuel tanks, 136 · Engine bearers, 136 · Stern tube, 139 · Cooling systems, 140 · Exhaust pipes, 141 · Fuel pipes, 143 · Fire precautions, 143

12
Miscellaneous, 145
Cockpit drains, 145 · Bilge pumps, 145 · Electronic gear, 146 · Navigation lights, 146 · Gratings, 146 · Binnacle, 148 · Sailing accoutrements, 148

13
Protection, Maintenance and Repair, 149
Protection, 149 · Maintenance, 151 · Repair, 152 · Minor repairs, 153 · Major repairs, 154 · Repairs afloat, 155

14
Raising the Wind, 156
Mortgages, 156 · Hire purchase, 157 · Bank loans and overdrafts, 158 · Tax relief, 158 · Acknowledgment, 159

Appendix I
Properties and Use of Materials, 160

Appendix II
Availability of Hulls, 165

Appendix III
List of Recommended Reading, 169

Index, 171

Foreword

There are practical problems connected with the purchase and fitting out of a hull moulded in glass-reinforced plastic (GRP, Fibreglass, Resinglass). In undertaking such a task myself I found little had been written on the subject. That seemed a good reason to set out my experiences for the benefit of others. Friends of mine have told me of their fitting-out problems, so that what is written in these pages is really a distillation of experiences. In addition, I have referred to what useful books and publications I could find and these are listed at the end.

I have made no attempt to teach such allied matters as joinery, welding, painting and varnishing, splicing and all the other arts and crafts that a Do-it-yourself boat-builder needs to master. There will be casual reference to butt joints, tenons and mortices, metal welds and so forth, but there is abundant literature about these things elsewhere.

What I have tried to give particular attention to are problems arising from the fact that resinglass (as I prefer to call it) is a relatively new material. This involves matters like bonding on wood and metal; making allowance for the mechanical properties of a material differing fundamentally from wood, metal and ferrocement—all used for boatbuilding. Resinglass, although tough, needs some protection and maintenance and this is mentioned.

It is my hope that readers will find this little book helpful to them and, above all, practical.

F. G.

I

Resinglass and Boats

A boat built on the foundation of a resinglass hull differs fundamentally from one of more traditional materials like wood and metal, although such materials may be used extensively in fitting out above and below decks. It will no longer be a conglomerate of many components held together with glue, screws, caulking, welding and so on, but virtually monolithic. This is because all its bits are bonded together, resulting in a detailed, composite but nonetheless integrated structure. Even if such a simple thing as a wooden batten is fixed to the hull it will not be an attachment but part of a greater whole. Its removal will no longer be a matter of removing a few screws and perhaps breaking a line of glue but, in reality, mean the destruction of a portion of the boat—an attack on the integrated boat/batten assembly. Luckily the resulting damage is simple to repair, as is also the case when accidental damage is incurred. Ease and cheapness of flawless repair is one of the virtues attaching to a resinglass hull.

Completion of one of these hulls demands different and more detailed forethought than would go into the same task with other materials. If the intention to build in say, a watertank, is overlooked in the planning stage it may be impossible to add it later because the space in which you wish to mould it has become inaccessible. Any item which has been incorrectly positioned, wrongly dimensioned or defectively constructed will have become part of the boat. Removal or modification may be difficult or even impossible.

A resinglass boat depends greatly for its strength and effectiveness on intelligent and informed planning followed by careful, conscientious workmanship and use of materials. The simple watertank, well moulded and properly sited, can act as a vital strength member in addition to holding water. Conversely, a massive looking bulkhead can be quite useless as a strength member if scant care and attention have been given to the method and means of fitting it in.

This dual role of primary function and structural advantage is not necessarily obvious to an untrained eye, but you will have to keep it in mind to get the best and most economical use out of your chosen materials. Waste incurred by careless planning can prove expensive because resinglass is not cheap and, unlike timber and metal, cannot be re-used successfully.

RESINGLASS

The material as commonly used is a cloth of one form or another made out of glass threads which, when saturated with resin, sets hard into a solid sheet (laminate). Most resinglass structures are fabricated by moulding a number of laminates together, the process known as lamination or laying-up. Although each layer could be allowed to set hard individually, each laminate is usually added to the lay-up while the preceding one is still wet or tacky. During lay-up it is usual to mould in strengthening items like timber stringers and such fitments as chainplates, deckeyes and stanchion sockets. This is stronger, neater, cheaper and easier than adding the items to a completed boat. It is one of the important considerations in designing a hull.

A good hull will have had the versatility of resinglass taken fully into account and builders will not have to add many items normally added to a wooden or metal hull. For example, a toerail can be moulded-in as part of the superstructure. Winch-pillars, adequately strong, can form part of the structure as moulded; so can thickened areas to receive cleats and other stressed fitments. Points to be considered when you are choosing a hull are dealt with later but it does no harm to emphasise, right from the start, that good design will save you time and money when completing, by realising the full potential of resinglass for hull moulding.

GLASS CLOTH

Glass cloth used for marine purposes is treated with an additive designed to integrate it with the resin so that it shall be, among other attributes, resistant to its environment. You are advised to use material of the correct specification. Bargain offers, bankrupt stock and similar attractively priced alternatives may mean that you end up with a boat which is unfit for its purpose.

Types of cloth include chopped strand mat (C S M) which is a rough cloth of random lengths (about two inches) of thick glass threads stuck together with an additive. As this dissolves quickly on wetting-out, C S M does not take kindly to overmuch manipulation or disturbance after being laid into place. Otherwise it is the easiest cloth to use and forms the staple of most bonding and simple moulding. Woven cloth (W C) is a loomed fabric and resembles cloths of other stuff like cotton and so on in having a warp and weft. It can be made of thick or thin threads, be loomed tight or loose, have its strength in one direction or another for different purposes and comes in various widths. Finished off with selvedges, it is neater to use than CSM and will take much more disturbance during wetting-out. It is considerably more expensive to buy and calls for more work and care in wetting-out. Woven rovings (WR) make up a very thick, stout cloth used where great bulk and strength are needed. Warp and weft are composed of bundles of thread, quite thick, and their compactness makes it difficult to wet it out. But unless that is done thoroughly the whole point of using the material for added strength will be lost. Lengths of rovings can be bought as such or plucked out of a piece of WR. They are used to lay along edges and to reinforce key structures moulded basically of weaker cloth. Woven tape (tape) comes in many widths from a half inch up to six inches and is selvedged on both sides. It will be as strong as WC of the same weave and can be used for both strength and a neat finish. It is relatively expensive and the same ends can often be achieved by using WC cut carefully and laid neatly.

Only two other types of glass material are likely to be used by an amateur; there are more and, if you think you need something for a special purpose, suppliers will be glad to give you details. Glass fabric is used to produce a thin laminate, usually found as sheathing over wooden hulls. Surfacing tissue is a gossamer cloth and used to impart a smooth finish to mouldings made with CSM or WR. There are coming on to the market finishing resins which, when used with tissue, ensure that the surface is not only smooth but resistant to abrasion.

I am not going to deal with carbon fibre in this text other than to remark on its properties. It confers unbelievable strength to a moulding in which it is used in place of, and in lesser degree to reinforce, the usual glass fibre. Regrettably it is prohibitively priced and the average boat owner will probably never have occasion to work with it. You might wish to buy a small quantity to experiment with—I have no personal acquaintance with it, but understand that it can usefully be employed in beefing up key points of a stressed moulding.

RESIN

Two main classes of resin are to be met with—polyester and epoxide. Polyester resin is the one generally referred to and used in resinglass work. Epoxy resin can be obtained in a number of types designed for special purposes like ultimate strength and adhesion, flexibility and so on but it costs about four times as much as polyester and is more demanding in its application. Still, I always keep a small supply on hand as it is useful for various purposes as described later on.

Polyester resin is a liquid chemical which will in the course of time automatically set solid (gel). The time during which it remains liquid, and therefore useable, is known as its shelf life. It can be used with success as long as it remains liquid, but toward the end of its shelf life it may become too viscous for easy working. The gelling process can be accelerated by mixing the resin with a catalyst and an accelerator. Commercially bought resins are usually pre-mixed with accelerating materials so that you do not have to add one yourself. Sometimes, however, resin may be delivered to which both catalyst and accelerator have to be added, so it is important to know what you are getting. Mixing a catalyst with an accelerator can cause an explosion. An accelerator must be mixed completely into the resin *before* any catalyst is allowed near; this mixing will not materially affect shelf life.

By varying the proportion of catalyst to resin the setting time of a mix can be controlled to fairly precise degrees. Most catalysts are basically peroxides and so have an inherent fire risk attached to them. They should be carefully mopped up if spilled and the rags, together with any empty containers, disposed of with care. Don't put them with flammable waste like woodshavings or paper.

The use of resins and catalysts has a risk of dermatitis which can be minimised if you use barrier creams on your hands, take care in handling them and scrupulously clean any contaminated skin with a molecular cleaner like Swarfega or Saroul. The risk is not great with modern materials, as manufacturers pay attention to this risk and are constantly improving their products. However, dermatitis is a nasty affliction and I take jolly good care not to run the risk while precautions are so easy.

Epoxy resin is also a liquid chemical but the catalyst has to be added in invariable proportions. Setting times can only be varied by controlling the temperature of materials and working space. Compared with a polyester mix, the setting time of epoxy is long—seldom less than twenty-four hours—which can give rise to difficulties of handling. Although the risk of explosion is absent, the same sort of other precautions should be taken. Large mixes of resin and catalyst are to be avoided as their reaction is strongly exothermic and can lead to spontaneous combustion.

POLYESTER MIXES

The setting time of a mix depends on many factors—proportion of catalyst to resin is the most critical, but ambient temperature has to be considered because setting time decreases with rising temperature. The reaction between catalyst and resin, as for epoxy mixes, is exothermic so that a quantity of mix will effectively raise its own temperature. A container of mix will gel faster than if the same quantity were spread out into a layer; a thick layer will gel quicker than a thin one. It is therefore sensible not to mix up too much catalyst and resin at one time. If inexperienced, you will find that a pint is about as much as you will be able to deal with in the beginning. Information about the physical properties of their resins and catalysts is usually given by suppliers as a matter of course. If you run into problems, they are generally most helpful in providing a solution.

It is essential to realise that catalysts are used as a very small percentage of a mix—typically somewhere between $\frac{1}{2}$ and $1\frac{1}{2}$ per cent—so that errors in measurement will have disproportionate effect on setting times. If you add too little catalyst a mix will take too long to gel, and this may hold up progress. Too much catalyst will cause the mix to go off before it has been wetted-in and this will expensively eat into stocks of both glass and resin.

As long as a mix remains in a truly liquid state, no matter how viscous, it is useable, but it goes off in stages which you should learn to recognise. If in any doubt, mix up a small quantity and look for the signs. It will be thrown away afterwards but may save you pounds later on. The first sign of gelling is the appearance of a skin or lumpy thickening in the liquid. The mixture is then useless for further application. It then progresses to the consistency of a syrupy jelly; a firm jelly; then a flexible solid. This flexibility means that the resin has reached the 'green' stage during which the laminate or moulding can be trimmed and pared with a sharp knife or chisel. The duration of the green stage depends on the speed of setting which, as I explained, depends on the catalyst/resin proportions and the temperature at the time. When a large item is being moulded which needs considerable trimming up and cleaning off when finished, it may pay you to use a slow-setting mix which will be followed by a correspondingly lengthy green stage. Thereafter the moulding will get progressively harder and more resistant to working, although for a considerable time it will be capable of permanent distortion if allowed to sag or settle out of shape. Eventually it will 'cure' to a shape incapable of permanent distortion, although it will always remain flexible within the limits imposed by its shape and thickness. The time taken for a moulding to cure fully could be of the order of weeks, or even months in really cold weather, so that you should really give immediate attention to a newly delivered hull. (See p. 47)

Quantities of resin are given as either weights or volumes; one pint of resin weighs 1½lb. As it is such a viscous and adherent liquid it is convenient to weigh out a quantity into a previously weighed vessel. If the same size of container is always used (I use a standard size of marmalade jar) one soon comes to be able to pour out a given weight by eye. Into this you will have to put the right proportion of catalyst. This has to be by weight, of course, but there is an alternative to attempting to weigh out very small amounts.

Catalysts come as either paste or liquid, and the latter is easier to measure. The weight which is contained in a filled eyedropper (price 3p) can be measured initially and used as the basis of all future operations. If, for example, a dropper-full consists of one hundred drops weighing one ounce in all, a dropper-full plus twenty drops would weigh 1.2oz, equivalent to one per cent when mixed with 7½lb of resin. For my standard jar containing one pint, or 1½lb, I know that I am adding one per cent of catalyst by squeezing out $120/5 = 24$

drops. I have scratched the glass of the dropper to show intervals of 25 drops, which is near enough. A hypodermic syringe is useful for this job as it is accurately calibrated. It can also be filled with the thin-paste type of catalyst. This, by the way, is usually prescribed by suppliers to be measured out in inches squeezed from the tube, so check up on how much is needed for various percentages. It will not be the same strength as a liquid, probably weaker.

TYPES OF POLYESTER RESIN

The resin is marketed in many forms—gel-coat, lay-up and finishing resins being the basic varieties. To any of these can be added substances which confer such properties as fire-resistance, flexibility, translucency and so on. Alternatively resins can be bought with the required properties embodied.

Gel-coat resin is used to give the familiar gleaming finish to the outside of mouldings and is invariably coloured. (I consider white as a colour for this purpose.) The colouring pigment can be bought and stirred in, which needs to be done very thoroughly to get an even finish, or ready-pigmented resin can be bought. Other resins can be similarly coloured but there is no point in using pigments for the inside laminations of a lay-up.

Gel-coat is applied generously to a prepared mould with a suitable tool; the minimum amount to be used is $1\frac{1}{4}$lb to the square yard. It will do no harm to exceed this quantity if the outside of the moulding may be vulnerable, but remember that gel-coat is brittle due to the absence of reinforcing glass cloth. Don't overdo it or you may wind up with a crazed surface at some later time. The coating contains a thixotropic agent which allows it to settle on sloping or vertical surfaces without sliding

FITTING OUT A MOULDED HULL

into puddles and leaving hollows and starved areas in its wake.

On top of the gel-coat are laid the requisite number and types of laminate which can be mixed with thixotropic or other media as desired. The specification of a lay-up should be given by a designer and might, for a hull, be such a combination as two layers of 1½oz C S M; one of 24oz W R; two more of 1½oz C S M and so on until the required complete lay-up has been achieved. Resinglass is very versatile, and you can lay it thick or thin according to the mechanical and other requirements of your moulding. A keel containing heavy ballast would obviously need to be thicker than a hatch cover; its base could be ¾-inch thick and the walls ½-inch tapering off to ¼-inch as the bilge is turned. It is not commercially viable to make such graduations in timber or metal, and these materials will be dimensioned to the maximum throughout.

It is possible to finish off the exposed interior of a moulding with tissue and finishing resin to give a fine, smooth surface. However, most hulls are left comparatively rough inside. This does not mean hairy, with fibres of glass exposed to the touch which would mean dubious wetting-out and is a point to watch for, but wavy and uneven to the eye. It would not be worth trying to achieve a finish like a gel-coat in an area where the surfaces will probably be covered over with paint, vinyl fabric, wood trim or other decoration.

USING RESINGLASS

Work with resinglass should be thoughtful, thorough and clean.

The main work you will be likely to encounter in fitting out will consist of bonding in bulkheads, stringers,

RESINGLASS AND BOATS

bunk fronts and the like to the inside of the hull. Attaching chainplates, cleats, tracking, deckeyes and so on to the deck is another chore. Then there is moulding or fabricating and attaching fitments like tanks, lockers, shelves, pipes, wires and many other items.

Resinglass objects are normally made by using either a male or a female mould, with a gel-coat inside or outside respectively. The type of mould employed will depend on the purpose of the moulding; a tank might call for a smooth interior and thus a male mould. A binocular box would likely have a gel-coat outside and a roughly finished interior to be covered with baize.

You can make moulds up out of a surprising variety of things—plywood, hardboard, resinglass itself, plaster for odd shapes, things like Tupperware foodboxes; even teacups. To stop the moulding sticking to the mould, some of the surfaces on to which the lay-up will go have to be treated with release agents. These are wax, of which many sorts can be used, or a polyvinyl alcohol liquid; often they are used one on top of another for various reasons. Glass can be used (I mean the ordinary, smashable stuff) to lay-up sheets of resinglass for cutting into strips and pieces to be used as described later. Polythene sheeting is another substance to which resin will not adhere and can be used to mould around a metal tube, for example, should you want to make a resinglass one of about the same size. If you mould on top of glass or polythene there is no need to use a release agent, but wipe them off with a dry, smooth cloth to remove dust before laying-up.

Porous surfaces like wood, hardboard, plaster and so on must have their pores sealed before the release agent is applied to them, and you can use something like Ronseal or a polyurethane varnish for the purpose. If moulding something up against the hull, say a water or

FITTING OUT A MOULDED HULL

fuel tank, it is perfectly in order to use the hull surface as one side of the moulding and butt the other parts of the mould up against it. Such a tank would have less material used in its construction and you would save time for the job.

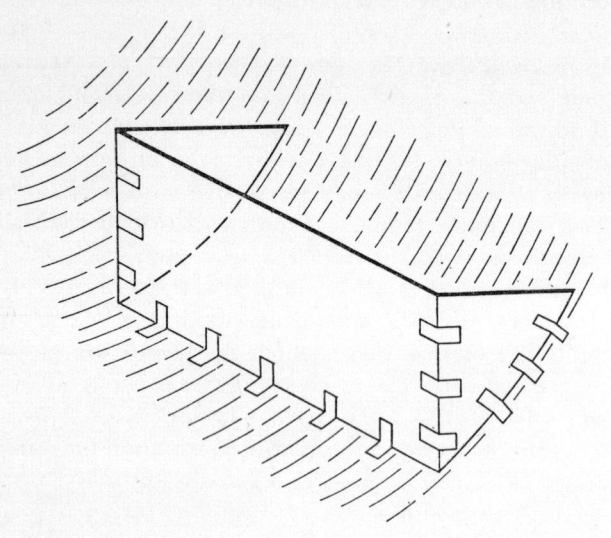

Tank mould stuck together and held in position against hull with strips of Sellotape.

Once a moulding has set you may have a little trouble in persuading it to come cleanly and easily away from its mould. If you have put the release agent on thoroughly and conscientiously, so that the moulding is not actually stuck to the mould, all that holds it in place is a vacuum. If this can be broken by prising gently at the edge of the mould with strips of hardboard (metal and wood may scratch the gel-coat), a moulding will often spring free with a disconcerting but innocuous 'crack'. A large moulding such as a hull needs considerable force to release it and to that end wood blocks or metal rings are often moulded in for attachment to a crane, derrick or other lifting gear. Anyone having release trouble with smaller mouldings could think about moulding in similar aids, but I find that a thump or two, carefully delivered, is fairly effective.

It is painfully obvious that a moulding should be of such a shape that it is physically able to come free when set. A taper in the wrong direction or underset flanges would mean destruction of the mould to release its contents.

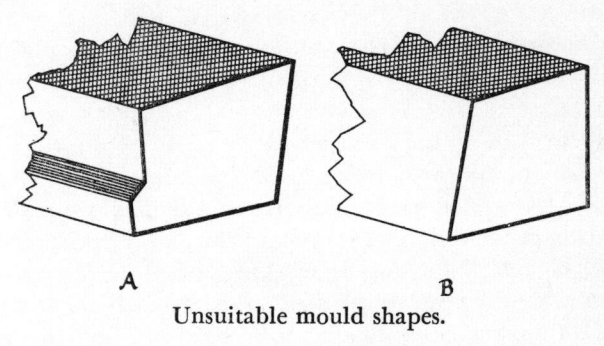

Unsuitable mould shapes.

A: Underset flange
B: Inward taper

This does not matter unless you intend to use the mould again. If so, it is perfectly feasible to make a mould in two or more sections which can be unbolted to release the moulding when completed and set. This is common practice with hulls which have, for instance, tumblehome to their topsides.

The most commonly used cloth, on account of economy as well as ease of handling and wetting-out, is probably 1½oz C S M. Two layers of this will serve for most bonding jobs. If you find that extra strength is needed it is simple to add another lamination or two. Appendix I indicates common usage. As experience grows you will find it pretty easy to estimate the requirements of different jobs, but a guide is handy.

Resinglass is only as good as the effort put into it. *Constant* attention must be given to proper wetting-out of the glass cloth. This means that you will have to stipple, roll, brush or otherwise persuade the liquid mix to penetrate every airspace between threads of glass so that there remain no voids, air bubbles or starved areas in the lay-up. These are all sources of weakness and, if not noticed before they are covered with succeeding laminations, are undetectable in a completed moulding.

A coat of resin is applied to the surface being laminated. The cloth is smoothly laid upon it and more resin worked in until the whole quantity appropriate to the weight of cloth being used has been applied and wetted-in. Appendix I gives details. Whether you call it wetting-in the resin or wetting-out the glass is immaterial, but the operation entails working away on the lay-up with your favoured implement until there is no sign of unsaturated glass. I like to use cheap shaving brushes which allow hard stippling; also, being lazy, I often throw a well-used one away rather than trouble to clean it. Other folk use brushes, paint rollers (look out for polystyrene ones which dissolve into the resin), special rollers with revolving metal washers—a whole host of tools is available and it does not matter which one uses as long as the end result is fully wetted-out laminations.

So that enough time can be devoted to conscientious wetting-out it is best to use a fairly slow-setting mix. One which you fear will go off at any moment may give rise to scamped and hurried work. It is better to allow for a little more time than you anticipate the job will take. As laminations are laid on in succession while preceding ones are still wet it does not greatly matter if the final setting time is reasonably slow.

BONDING TO CURED SURFACES

A hull will arrive some, if not all, of the way to being fully cured. It helps if the date of moulding can be obtained from the supplier, for a new moulding will need to be shored and chocked up quickly and accurately on delivery. Otherwise it could take up a distorted set. However, no matter at what age it is received it will not be practicable to bond things to it really successfully without pretreatment. Any surface to accept a bond must be free of grease, dirt, paint and other contaminants. Grease-removing fluids are easily obtainable. After using them, which entails little more than spreading them with a paintbrush, the cleaned surface should be washed off thoroughly with fresh water and allowed to dry completely. Any trace of moisture will adversely affect bonding.

The prepared surface should then be abraded so that fibres of glass are brought up to stand proud of the surrounding resinglass. Sandpaper, file tangs and things like hacksaws and knifeblades are of little avail. I use a strip of $\frac{1}{2}$-inch plywood into which I hammered a close pattern of a couple of dozen masonry nails. These keep very sharp and tear deep into resinglass surfaces. Use such a thing brutally; the resulting scars will vanish into the bond and have beneficial effects on the operation.

Infrequently you may have to bond over a gel-coat, and for best effects this should be chiselled away so that you can get at the underlying fibres for roughing up. Use a sharp wood chisel, but strike it gently in a horizontal plane so that the cuts do not penetrate too deeply into the laminations.

FITTING OUT A MOULDED HULL

An added precaution to good bonding is then to brush the torn area with acetone or styrene. Either fluid helps slightly to soften the cured resinglass and assist the chemical bond between old and new materials. Styrene is more expensive but, as it is in any case a constituent of the resin, combines chemically. Although perfectly compatible with it, acetone does not combine with resin and should either be wiped off or allowed to evaporate almost completely before the bond is made.

Take care not to spill either liquid on to surrounding areas as they will leave indelible marks; puddles allowed to stand will tend to dissolve and weaken the structure. To prevent damage it is advisable to mark out areas of work and outline them with wide masking tape or special resinproof sticky tape. If you peel the tape off carefully just as the bond goes green, a clean line will be left and you will be commended for professional looking work. Nothing seems so amateurish as a wavering line of resinglass bonding; it is almost impossible to clean it up or trim the edges once it has set.

BONDING TO OVERHEAD SURFACES

If you try to apply resin followed by cloth and more resin to the underside of a horizontal surface, much splashing and dripping is almost inevitable. The resin should have pre-gel or other thixotropic material mixed in—proportions are stipulated by suppliers. This ensures maximum resistance to slipping.

First apply a coating of resin to the area being treated, in the normal way, but make it thinner than normal so that it does not sag.

You will then need a piece of prepared hardboard or plywood—I find that it is best to cover it with polythene sheeting stuck on the back with sticky tape. On this lay your first strip of cloth and wet it out thoroughly. Lift it up into position and, with careful wangling, slide the piece of board away from it. The wetted cloth will then be roughly in position and can be moved gently into place with the wetting tool. By repeating the operation, you will be able to complete the bonding with an absolute minimum of mess. It helps, in such a case, to have a quick-setting mix so that the laminations go off before gravity can cause any serious sagging. It may be necessary to keep prodding at the cloth during the period of setting to push back bulges and slackened edges.

BONDING TO WOOD

Again, good preparation is half the battle. Due to manufacturing processes, plywood does not present a good surface for bonding. It has to be thoroughly scored and rasped until wood fibres stand proud, and then given a coat of styrene-thinned resin (1 styrene : 5 resin) which should be brushed well into the abraded area and left to go off hard. Bonding made within a few days after that will be excellent. You should treat planed or smoothed wood in similar fashion, but unprepared timber bonded in for strength will be rough enough to accept the thinned coat without further ado. Ply or any other sort of wood may have been treated against rot, or contain contaminants most of which have a deleterious effect on resin. Use untreated wood, if possible, and remember that any wood should be free of damp.

BONDING TO METAL

Grease and gloss are enemies of good bonding and all metal should be regarded as greasy even if nothing is visible. Surfaces should be roughed up with a file or coarse sander and then given a de-greasing with methylated spirit, trichlorethylene or other similar substance. Having gone to this trouble, take care in handling the prepared surfaces; the human hand is naturally greasy. (Some, like my own, often unnaturally!) When incorporating plates, or other pieces of metal, into a lay-up or a bond it helps to bore plenty of holes through (without unduly weakening the metal) so that the resin percolates through from side to side before it sets.

Some metals, notably those containing copper, affect the setting reaction of polyester resins. This is not so with epoxies, which should therefore be used to precoat yellow metal fittings, copper piping and so on. Wet epoxy resin should not be combined with wet polyester, so you will have to allow time for the coating to set hard before getting on with the bonding.

USE OF CLOTH FOR BONDING

Details are given where necessary in the text, but the following notes have a general application.

Resinglass, in common with many non-malleable

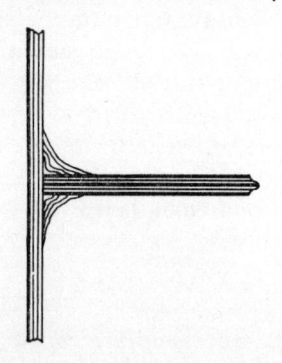

Bulkhead butted close against hull—a hazardous practice.

materials, is relatively weak in the bending mode. This means that bending stresses should be spread over as large a radius as possible; the smaller the radius the bigger the risk of fracture. If a resinglass laminate is bent around an edge, or worse still around a point, stress is severe. The drawing shows a bulkhead butted to a hull.

A shock stress, such as caused by falling off a wave, would cause a situation which is basically bending part of the hull along a line.

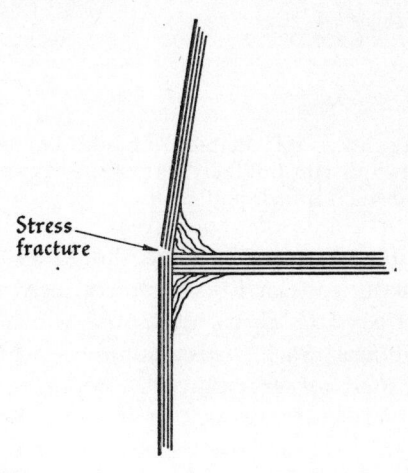

Line fracture caused through butting bulkhead against hull.

The remedy is to bond in such a way that the radius of curvature is increased to a tolerable dimension. Hard edges, like those of bulkheads, should be spaced a small distance away from the hull and not butted directly against it. You can also put a softer material between edge and hull—balsa wood, neoprene or something with a bit of give to it. In any case, whether the space is filled or not, the bonding lamination should be graduated so that the maximum thickness is at the curve of the bond and the laminations taper away as shown.

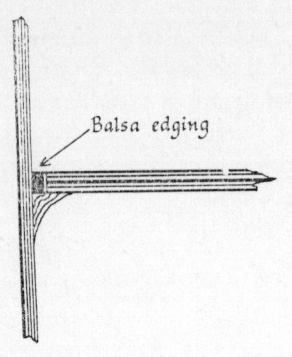

Balsa or other soft material should be interposed between hull and bulkhead. Alternatively an air gap of about half an inch will suffice.

The principle of spreading loads should also be applied to any backing pads of wood or metal used to reinforce stressed fittings like cleats, deck-eyes, winches, fairleads and chainplates. Such pads should be chamfered all round and their corners radiused.

If you thicken up the hull so that you have in effect added a pad of resinglass, remember that this should similarly be chamfered and radiused. If you leave it with perpendicular edges, lines of stress will be created and this is just what needs to be avoided in such places.

Anything with a selvedge, like glass tape, is easy to handle but some cloths, particularly C S M and W R, need to be handled carefully when you are cutting them into shape. I find it best to put them on a flat wooden surface and to lay a lath or batten along the proposed line of cut. I hold it down securely and draw a sharp edge carefully and often enough along the line of cut to ensure complete severance of the piece from the bulk. A Stanley, or similar knife is fine for the job, but I have learned not to use wood chisels; the glass plays havoc with their edges. Cut pieces are then lifted away and carefully laid flat for use. If you lay them slightly askew from one another in piles they will be easy to pick up individually;

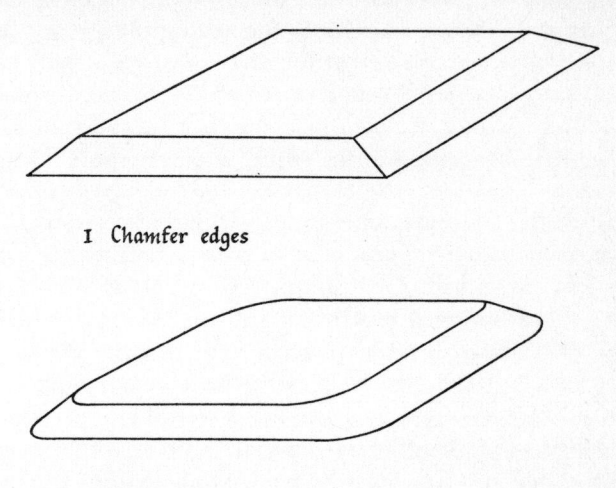

I Chamfer edges

II Radius corners

The proper way to make backing pads.

there is a tendency for projecting fibres to intertwine and the pieces stick together. Accurately sized pieces are not needed as will be told later, but take care not to handle such off-cuts unnecessarily as they may start to fall to bits along the edges. Woven rovings are very prone to shed long wefts. Glass fibres are hard and sharp and you will do well to take my tip and beware of splinters. They dig in and hurt worse than do wood splinters; if troubled you can wear plastic gloves. All cut pieces should be kept dry and well away from possible contamination by grease or dirt.

OTHER METHODS OF ATTACHMENT

Sometimes bonding is neither convenient nor feasible and you have to turn to more traditional methods of attachment—glue, screws, etc. A nail driven into resinglass will not hold and will only create a useless, unsightly hole. Screws, but only the self-tapping type, can be used to a limited extent for direct fixing but will not take much weight. Woodscrews can be passed through resinglass into a wooden backing pad or part of the structure, say a bulkhead, using a counterbore as for wood.

Instead of screws, and usually with better results, it is easy to use an epoxy glue such as Araldite. If large areas are involved, for example fixing slats to a cockpit sole, or battening a coachroof, you can use epoxy resin instead of the glue which is more expensive for the same quantity. Neither are cheap but a little goes a long way and nothing else will provide such a secure job. Problems arise due to the lengthy setting times, during which glued or resinned surfaces must be held immobile and in close contact.* This you may find tedious if you cannot get a clamp around the work or if it is on sloping or vertical surfaces. Screws will hold the components together until everything has set, but the strength of the fixing will be in the glue, not the screws, and you can remove them if you like. Holes can be filled with matching plugs for the sake of appearance.

Alternatively you can construct some light shoring or use Spanish windlasses of light line. Such little problems are part and parcel of boat-building and their resolution a source of self-gratification to any inventive person.

* Quick-setting epoxy glues have very recently come on to the market under various trade names (e.g. *Bostik*), but their suitability for marine applications has still, in my view, to be satisfactorily proven.

Using a Spanish windlass for light shoring.

SANDWICH CONSTRUCTION

Many decks, and a few complete hulls, are made of a sandwich with resinglass as the bread and other stuff as the meat. This could be endgrain blocks of balsa, plywood, polystyrene or other convenient material. Special care has to be taken in attaching things to sandwich structures as the resinglass skin is relatively thin on both sides, and a bond made to either one may prove unsatisfactory. Applied stresses in an unfavourable direction will tend to separate the layers of the sandwich and the area of attachment might bulge, crack or tear.

It is best to use backing plates for fittings and, if the load may be an alternating one, to fit such plates on each side of the sandwich. The plates can be bonded to the resinglass and it is an added precaution to fill any holes bored through with a spacer between the shank of the bolt used and the sides of the hole, as shown.

This will help to prevent sheer stresses causing the bolt to cut into the thin resinglass; the filling may be unresistant to them.

FITTING OUT A MOULDED HULL

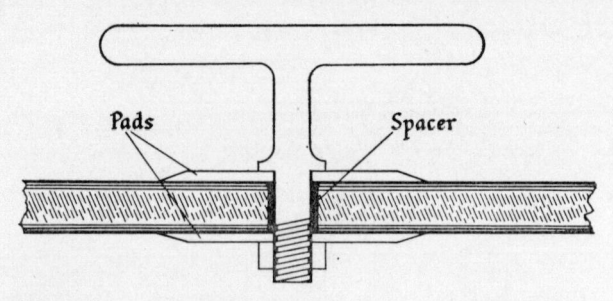

Mounting a stressed fitting through a sandwich hull.

If you have occasion to bore through a sandwich for any reason, it is wise to seal the sides of it so that moisture cannot penetrate the interior of the sandwich and percolate between its layers. It may well have no egress and so give rise to rot or freezing troubles. In fact, aside from sandwich construction, it is as well to seal all raw edges made in resinglass with a lick of catalysed resin. This guards against moisture creeping, under capillary action, along any fibres which may have escaped being completely wetted-out. There is no danger of rot, but moisture introduced into laminations in such a manner will stay there indefinitely; periodic freezing and thawing may cause delamination.

TREATMENT OF CURED RESINGLASS

In completing a bare hull it is inevitable that you will have to cut away, trim, bore into and otherwise impair its integrity in order to add to or modify its structure. Think of resinglass as a weak metal, like aluminium, and not as a superior sort of wood. Tools should be chosen and used accordingly. Admittedly, woodsaws will saw through resinglass and wood bits bore through it. They will, however, blunt quickly and leave ragged edges needing trimming up and cleaning off. It is better to use metal-drilling bits, hacksaws and other tools designed for metal and hard plastics.

Sanders, preferably orbital, can carry coarse paper to reduce edges rapidly, but these should be finished off with medium wet and dry paper, used dry. An Abramesh or similar openwork sanding disc is very effective on resinglass as it does not clog easily, but extreme care must be taken to control its travel if accidental scarring is to be avoided. If I could sound a warning note here; these hard cutting discs are, in fact, a form of circular sawblade and are completely unguarded. You will be wise never to lock the trigger of the electric drill being used, but keep a finger on it ready for instant release. If a disc catches in anything and pulls out of your hand, it can cause havoc if still running. Danger to your person could be serious.

Another danger, often discounted, is the real one of inhaling resinglass dust, and you really should take the trouble to wear a simple breathing mask when sanding down and sawing resinglass. Even in small amounts the dust can give rise to nasty things like silicosis.

To impart a polished finish to cured resinglass, wet and dry paper of fine grade should be used wet, plenty of water being applied continuously and copiously to wash away detritus.

Drilling is best done at low speed and you should always enter into any gel-coat side to prevent chipping. If using an electric drill you may get the best results by using an electronic speed controller to slow down the rate of drilling without loss of power. On emergence a drill bit tends to leave straggly ends of glass around the hole and they will have to be sanded away. A backing pad of scrap wood will greatly reduce this tendency.

For cleaning up and reducing resinglass evenly, a

Surform is very effective, used with a blade designed for metal and plastics work.

It is often necessary to scribe a surface before cutting, and wood markers are too soft to leave a clean line and also wear down rapidly. A metal scribing tool is to be preferred. Circles can be marked with ordinary drawing dividers which have surprisingly hard points and stand up well to such usage.

CLEANLINESS IN WORKING

A resin mix will spill, splash and trickle like any other liquid. Having done so, it will then start to gel on the recipient surface. If this happens to be cured resinglass removal will be difficult and a stain unavoidable. Precautions are essential in the interests of clean and undefaced surroundings.

Polythene sheeting is quite cheap and should be spread well around areas surrounding work in progress. It will not matter how much resin falls on it unless you put your tools or hands into it, or carelessly kneel in it. Pieces of sheeting can be cut to fill awkward spaces and tacked into position with sticky tape; this will come away without leaving a mark when the sheet is carefully removed. It is probably best to leave a bespattered sheet in place until splashed resin has set hard.

A visit to a moulding works will show you that people engaged on laying-up work with one hand behind their backs. With the other, they carefully dip the roller, or other tool, into the container of mix, wipe the surplus off against the inside edge and then apply the loaded implement to the work in hand without haste or flurry. At the end of a stint an operator's hands will be clean, and so will the handles of his tools. We should all strive to emulate such professionals, and with reasonable care and practice hands and tools can, indeed, remain uncontaminated. However, accidents happen and wayward resin has to be cleaned up.

Resin landing on an unprotected surface should be carefully mopped up with a *dry* cloth. Wipe inwards from the edges until the bulk has been removed so that it is not spread any further around. The remaining smear can be persuaded away with a smooth cloth soaked in acetone or styrene. These softening agents should be used very carefully as, if you brandish your cloth too freely, they will turn a small stain into an area of defaced finish. If time permits, and drops or splashes are not near going off, you may find it helpful to surround them with sticky tape before starting to clean them up.

It is a good habit to rub plenty of barrier cream into your hands before you start work. This stops the resin getting a grip on the skin, and any that lands there will come away fairly easily with a suitable hand cleaner.

Resin on clothes is there to stay. The only thing to do is to assume the worst will happen and wear absolute rags when working with it. If you are unlucky enough to get wet resin on your clothes, take care that it does not rub off on surrounding surfaces.

A realistic way of avoiding accidental spills and drips is to mix the resin in a container which is then put well down inside a plastic bucket. This is noticeable and less likely to be kicked over than a small jar. Resin spilling from the container into the bucket can be left to set and you can remove it by flexing the bucket. Digressively, containers should never be of polystyrene as this substance is dissolved by resin. Waxed containers, like yoghourt cups, are perfectly all right to use.

Tools used for applying a mix should be kept clean during the operation, and it is not advisable to use one for so long that resin begins to gel on it. This means that you will have to alternate two or more tools during the life of a mix. Those coming out of use should be immersed in a container of cleaning fluid; this is usually acetone although I have heard of neat detergent liquid being used instead. I imagine that there might be trouble if this got into the resin later. Acetone is relatively cheap but it is flammable and fire needs to be guarded against. There is a non-flammable cleaner which I have used, known as GRP Resin Solvent; some people dislike its odour but I did not find it objectionable. Current prices, agents and so on can be obtained from the manufacturers—Wilcot (Parent) Co. Ltd., Fishponds, Bristol BS16 2BQ. Other cleaners could be styrene, which is expensive, or cellulose thinners, which might not be entirely suitable.

The sketch shows a practical method of keeping brushes and rollers ready for use—the resin loosens from the bristles and drops down to the bottom of the container. If they are just dropped into a container of

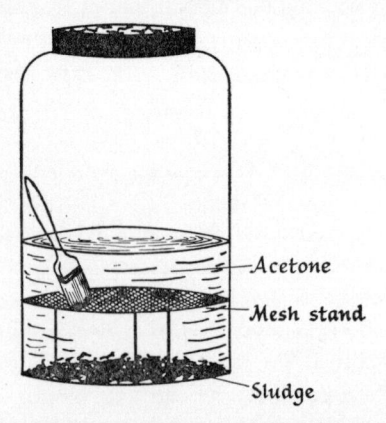

A wide-mouthed jar for keeping brushes clean. Crumpled chicken wire will serve instead of the mesh screen.

cleaner, a sludge forms on the bottom and gums them up. Any traces of resin remaining can be flicked off together with surplus cleaner when a brush is brought out for re-use. It is not necessary to wipe the implement off further than this, but it is obviously undesirable to introduce acetone or other substance into the mix in appreciable amounts. Brushes and rollers being stored later should be washed out with detergent, rinsed and dried in the normal way.

DECORATION

As I have explained, resinglass is usually coloured by pigmenting the gel-coat and interior surfaces of the hull. It may not always be possible to use pigment to colour an attachment or fitting, but it is simple to paint resinglass if a routine is followed. Used and repaired hulls are very often completely painted over to cover fading and remoulding, so that there is no reason to think of painting new resinglass as anything extraordinary.

Paint will not adhere well to a gel-coat or other cured surface unless they are specially prepared. First, a grease remover is applied and washed off. Once dry, the entire area to be painted has to be roughened either mechanically or chemically. Even with an electric sander, the task of abrading a large area with wet and dry, used dry, is laborious. You may find it easier to use chemicals. An etching fluid, of which there are many brands suitable for resinglass application, is applied to the degreased surface and left to react for the recommended time. Washing down may then be prescribed, depending on the brand used, before the hull is dried off and paint put on.

There are many kinds of paint which can be used with varying degrees of success and durability. Ordinary gloss paint gives an acceptable finish but usually needs an undercoat or two to get a good effect. It does not stand up well to marine conditions and should, if used, be restricted to interiors. For exterior work you can choose between acrylics, marine enamels and 'yacht paints', and one- or two-part polyurethane and epoxide paints. Two-part epoxide paint is really a form of epoxy resin and most durable, but it is difficult to use and results can be disappointing. It dries quickly to an initial set so that a 'wet edge' is hard to maintain. Despite this it can be affected by damp conditions and may become fogged or mottled unless you can rely on a period of warm, dry weather when using it. If things go wrong, it is very difficult to remove. However, a successful application will give you a brilliant and lasting finish capable of taking rough usage without undue defacement.

Two-part polyurethane similarly sets hard and is durable; it suffers to a lesser degree from the disadvantages of the epoxides. One-part polyurethane is not so hard, but it is quite durable and, on balance, I use it myself as probably the best paint for a hull exterior. Acrylic paint gives a nice finish and is easy to use, but I have seen a few examples which did not seem to have stood up very well to weathering. There are new paints coming on to the market all the time, for example the vinyls, but I have no experience of them.

Interior decoration is not only a matter of painting; you can get involved with vinyl fabric (which is sometimes backed with foam), polystyrene tiles or sheets, thin plywood, tongue and grooved timber laid over battens, and many other things. Most of these will need to be stuck to the resinglass of the interior and, apart from wood which can be bonded on in the normal way, there is no call for any great strength in the adhesive used. Different materials are recommended to be stuck on with specified adhesives—Evostik, Bostik and the like—all of which may be expected to stick well to resinglass which has been roughened with coarse sandpaper. The surface need not be so rough as for bonding.

FIRE RISK

Resin—raw, mixed, or set and cured—is flammable. Once alight it burns fiercely and gives off a heavy, acrid smoke. It is difficult to extinguish.

When working with resin and in and around resinglass structures you should have means of extinguishing fire ready to hand. It is only sensible to take every precaution open to you to prevent an outbreak of fire.

Naked flames and electric leads should be kept well away from work, and the use of matches and tobacco watched carefully. Electric bulbs and other forms of heating are often used to speed up setting times in cold weather, and you may wish to warm up your surroundings when working. No heating device should be left unwatched; it is simply asking for trouble to go away and leave a fire or string of bulbs alive inside a hull. Catalytic heaters are effective and pretty safe if used correctly, but bear in mind that the use of propane or butane gas is accompanied by a risk of explosion.

Prevention of fire in relation to fitting out is contained in the text where appropriate.

2

Deciding on a Hull

Labour costs account for a large part of the price of a commercially completed and fitted-out boat. By eliminating them a bigger or better craft can be owner-built for the same sum of money. Apart from financial considerations there is a pride and satisfaction in building your own boat. She can be individual in design and construction, and you will have a precise knowledge of the materials and care that have gone into her making.

Unless merely adding in a manufacturer's kit of parts and equipment, an aspirant builder will have to study the design of rig, fittings, accommodation and much else. He will have to become familiar with the purpose and use of tools, properties and handling of materials, and to delve quite deeply into the general field of sailing (which term does not exclude powerboating) and allied matters. Such study, allied to possibly already wide experience of sailing, will enable most enthusiasts to produce a boat which conforms to their own current ideas of perfection. It is feasible to introduce innovations and experimental features, but for the most part it may be wiser to stay well within the bounds of tradition and convention.

Taking the matter to a logical end, you may ask why the hull itself could not be moulded at home, thus making further considerable savings and, maybe, resulting in a better designed and more soundly constructed one than could be bought. It is a good question, but there are a lot of factors involved which are not altogether obvious, although it is true that good hulls have been both designed and built by amateurs. Quite a few well-known names in the boat-building business belong to men who started out with the intention of building a dinghy for the children. The field is no longer quite so wide open as it was in the earliest days of resinglass.

The first snag is that you will virtually have to build three hulls to finish up with the one you want. There is a fairly standard process involved which is something like this:

HULL MOULDING

To make an acceptable resinglass hull it is necessary to build a full-scale model. This is usually of timber which admittedly may be of inexpensive softwood, even scrap timber, but the costs of completing it for useable purposes far outweigh the cost of the materials used. This plug, as it is called, has to have an immaculately faired off and polished exterior surface. Any imperfections in the plug will be faithfully reproduced in the finished hull, so that an immense amount of time and trouble needs to be expended on its construction.

Next, gel-coat and laminations are laid-up on the plug and, when set, the resulting structure is strengthened with timber, metal and as much extra resinglass as is needed to make sure that it will be absolutely rigid when taken off the plug. This makes a female mould, usually referred to as 'the mould', from which the final hull moulding will be taken. Due again to time spent and large quantities of material used, the mould is an expensive item.

Thus, if you employed this method of turning out a single hull, you would end up with an expensive plug and mould which were of no further use, and whose scrap value would be little.

These costs would have to be added to the price of time and materials incurred in making the hull moulding, and this would be quite excessive compared with the same hull produced in traditional materials like wood or metal. On the other hand, if many hulls are taken from the same mould, the initial costs of plug and mould can be apportioned between them, so keeping down the price of any one hull to a viable level. It is clear that this is a strictly commercial proposition, not open to the amateur who only wants one hull. Even assuming that he were prepared to stand for the exorbitant costs, for the sake of getting what he really wanted, other snags arise.

DESIGN

Reputable moulders just cannot afford to produce unsaleable wares and customarily employ the services of professional designers who can be relied on to turn out drawings which will ensure the production of seaworthy and vice-free boats. Their services come high, but series production usually results in a reduction in design costs for any single hull. It is no extravagance to use the best designers, for their art is esoteric. Any amateur lacking sound knowledge in the field and a flair for design is extremely unlikely to be able to turn out the satisfactory lines without which a hull may not be acceptable. Good hulls have come from amateur boards but hull design is no simple matter.

MOULDING

Even armed with the best of designs I would hesitate to mould a hull myself. Not because of unwillingness to undertake an onerous task, but because a hull is a vital structure in which weaknesses may be inbuilt for a number of reasons. Temperature and humidity control are needed to determine accurate and predictable setting times; precise measurement of large quantities of resin and catalyst so that all mixes shall be consistent; accurate measurement of additives, principally pigment, so that gel-coats shall be homogeneous; and more such technical stuff. As I said before, wetting-out is tedious and critical; it takes a practised eye to discern faults accumulating during the process of laying-up. Construction aside, the physical handling and transport of plug, heavy mould and moulded hull call for skilled hands and probably some mechanisation.

All in all it is cheaper, safer and less laborious to buy a bare bull from the experts in the field of moulding.

AVAILABILITY

A bare hull is very bare indeed. If you are thinking about a home completion it may help you to understand what alternative supplies are available. Some firms supply bare mouldings and it is pointless to ask them even to fit elementary strengthening bulkheads, rubbing strakes and the like. By concentrating on a narrow field of production they are able to keep prices down to a minimum. Other firms supply not only hulls—which they either mould themselves or have made for them under a sub-contract by specialist moulders—but almost everything else needed to fit-out and equip a craft ready to launch. The market is always changing as firms alter their policy, or arise and go under, and Appendix II lists a number of better known firms from whom details of their products can be got on request. Due to ever-changing conditions it would not be of much use to give anything of a detailed list in these pages.

Short of buying a fully completed and equipped boat, you are able to buy:

Bare hull.
Superstructure.
Sub-mouldings, or modules, like galley, lockers, bunks, bulkheads, cockpit wells or floors, hatch covers, tanks, driptrays.
Fittings, spars, rigging, windows, trim.
Fabrications of wood and metal such as bulkheads, stringers, engine bearers, window frames, lockers.
Engines and stern gear.
Kits of parts either of mouldings or fabrications (or both) which can include some or all of the above items and many more not mentioned. The permutations are infinite.

Most items can be bought individually and may be supplied loose or bonded-in. Hull and superstructure can be delivered separately or bonded together, and so on. You may, of course, decide on a hull for which a kit, or ancillary parts and fittings, is not available. For this reason it is best to study the market with a view to choosing a hull which will be acceptable not only for its intrinsic qualities, but for convenience of construction.

Labour costs are bound up in the price of any parts supplied ready-made or bonded-in so that it will invariably be cheaper to buy a bare hull, or hull and superstructure, and complete it by buying your own raw materials for the purpose. However, I must not be dogmatic and the matter is for fine judgment. Bulk production of, in particular, kits of parts enables firms to buy raw materials to much better advantage than a casual purchaser, so that the apparent advantage of buying and using raw materials may not be a real one. Questions of feasibility and personal limitations also intrude, and it might be more advisable to buy larger and more critical items inclusively with the initial purchase. For instance, if you thought of having a wooden deck and superstructure, it would be a demanding and laborious task to fashion and bond in gunwales and deck beams, perhaps not compensated for by the marginal savings achieved by doing it yourself. In the final analysis it boils down to the balancing of cash against time and trouble.

HANDLING A NEW HULL

Green mouldings are horribly floppy and you will have to avoid any possibility of a hull setting out of shape. This means the immediate addition of rubbing strakes and key bulkheads. If at all possible, regardless of the small extra costs, it is best to buy a recently moulded hull with such features already incorporated. Hull and superstructure bonded together are much more rigid, but you must take care to shore them up so that their weight is taken as evenly as possible all round the bottom of the hull. If you put all the weight on the keel it will inevitably result in a certain amount of distortion.

Suppliers should be approached about plans and drawings so that you can set about designing accommodation and fitting-out to your own ends. If you cannot get drawings, it may prove frustratingly wasteful of time to take off lines and measurements.

FITTING OUT A MOULDED HULL

PURPOSE

Choice of hull will depend primarily on the uses to which you wish to put your boat. Experienced sailors may know precisely what they are looking for, but novices in the field should ponder long before buying. A boat has many interdependent aspects, the main ones being seaworthiness, handling characteristics, habitability, appearance and, importantly, price and value which are not quite the same thing.

SEAWORTHINESS

Irrespective of all other considerations it would be folly to choose an unseaworthy hull for completion. A boat must be fit and well-found for her allotted tasks. A good estuary cruiser may be unfit to go out to sea; a coastal cruiser would be out of place in mid-ocean; and so on. When afloat in her correct environment she should be able to accept and deal with all conditions likely to be met. In all cases integrity of materials and workmanship, fittings, rigging, spars and other equipment must be beyond doubt. A boat must carry all the gear needed to ensure the safety of her crew, and be able to look after them when in straits. The best way of ensuring that your purchase is going to be basically seaworthy is to buy the products of good designers and reputable moulders. The initial costs will naturally be higher than if a boat of dubious origin were bought, but it will be more sensible to consider buying a smaller, reliably made craft.

HANDLING CHARACTERISTICS

A cruising boat, in particular, should be docile and predictable in behaviour. It is not easy to deduce her

DECIDING ON A HULL

potential from an inspection of plans or hull and, again, you can only rely on the repute of designer and moulder, the latter being responsible for translating the designer's intentions into concrete form. No boat, unless of a well-proven class, should be ordered without a trial sail or two and even with well-known designs it may be that what suits one owner will be unacceptable to another. Sailing boats with auxiliary power need to be tested under both methods of propulsion in hard conditions, for it is then that one or other might be found lacking.

PERFORMANCE

A boat should perform well within her type. A fast racing boat may have vices which would make her quite unsuitable for comfortable cruising, but acceptable because of her speed and other racing qualities. Conversely, an otherwise nice cruiser might be much clumsier to handle than others of her breed. Again, this can be found out either by repute or taking trial sails.

HABITABILITY

A cruising boat must be fit to *live* aboard. Things which make life less than civilised, or even tolerable, are inconvenient accommodation; lack of headroom and other space to move about in; inadequate lockerage and stowage; skimpy side-decks and cockpit space; inability to stow a tender on deck; poor or no ventilation arrangements. There are many other such shortcomings and a lot of them are caused through trying to crowd too many berths into a hull. Some small measure of overcrowding may be acceptable in a boat which only makes short passages, but if a crew is to live decently at sea for any period overcrowding is to be avoided at all costs. In an attempt to provide sufficient living space the purchase of a large hull may prove expensive, as price rises in proportion to the volume of a boat, not its length. This consideration applies to completion costs as well as initial ones. Much care is needed to design accommodation in an economical way, and good designers will plan things better than an amateur could. An attempt to modify an original design simply to pack in more bodies is not advisable.

APPEARANCE

The aesthetic appearance of a boat stems from the shape and congruity of her hull, topsides, rigging and fittings. An ugly boat *might* be perfectly seaworthy, and I can think of a few, but it is likely that her appearance is due to a basic disproportion brought about by an effort to overcrowd her interior. This may well cause undesirable characteristics. In the extreme such a boat could be positively unfit for her purposes.

Aesthetics apart, the appearance of a boat is quite often a clue to what you can expect of her. Racing craft tend to a low profile, with resulting lack of headroom, and indicate wet decks and excessive angles of heel caused by tall masts and great areas of high aspect ratio sails. Out-and-out cruisers will look chunky and have underwater lines and profiles implying sedate progress and slow response to the rudder. It is not really feasible to combine the requirements of racing and cruising in the same hull with complete success in either field. You are advised to be quite sure of your intentions before ordering a hull. People buying a pure cruising boat are often inveigled into and become interested in club racing; they wish they had bought something with more speed built into it. Those buying with the intention of racing

DECIDING ON A HULL

as well as cruising may get fed up with pottering round the local circuit and go over wholly to cruising. The matter needs considerable thought.

RIG

A good hull can probably be successfully rigged in a number of ways—bermudian, gaff, gunter, sloop, cutter, ketch, yawl, schooner and so on. The standard is masthead bermudian—a handy, simple and boring rig—and you may wish to have something a little more distinctive. This brings in the question of performance and appearance. You should read about the advantages and disadvantages of different rigs to help you choose a sensibly compatible hull. There is a good chapter on the subject in *Offshore* by Captain John Illingworth which is worth studying, although he is primarily concerned with racing matters. Scale drawings of different rigs set on the same hull make a comparison of appearance easy.

POWER

You can choose between inboard, outboard and the clumsily named inboard/outboard engines which can be fuelled with petrol, paraffin, vapourising oil or diesel fuel. Stern gear varies from directly driven two- or three-bladed propeller, variable-pitch propeller, folding and feathering ones: jets: forward, neutral and reverse gearboxes, reduction gears, hydraulic transmission. A hull must be chosen with regard to its ability to accept the means of propulsion preferred. If, for instance, you intend to fit an inboard engine after your boat has been launched, the boat as completed at that stage should be able to accept the gear without undue difficulty of installation. Unless planning has allowed for this eventuality you could be faced with extensive and expensive modifications.

It is quite feasible for a builder to equip his boat with a DIY engine instead of a commercial model. Much money can be saved and good results achieved. There are books on the subject, and I recommend *Marine Conversions* by Nigel Warren. For those of practical bent, he writes clearly and frequently on this and allied matters in the magazine *Practical Boat Owner*.

PRICE

It is vital to restrict the initial price of the hull to allow for purchase of all else needed to fit her for sailing—raw materials, tools, spars, sails, rigging, deck gear, ground tackle, navigational equipment, domestic gear and much else—a formidable list which you might well compile and cost out beforehand. Things often overlooked in the first flush of enthusiasm are delivery charges, cost of renting a building site, travelling costs during building, lifting and launching costs, insurance while completing, shelter or covers, a building cradle, staging or trailer. And you will not be likely to overlook the effect of inflation on all these things during the period of building, which could be lengthy.

The question of size has been touched on, but having thought about the above points you may find it better to settle for what will be a well-found and comfortably fitted boat somewhat smaller than your dream. Too much money spent on a commodious hull could find you landing up with a scantily equipped source of frustration and vexation.

DECIDING ON A HULL

VALUE

Do not overlook resale value when buying your hull. It seems that, surprisingly, people change their boats about every three years no matter how certain they are, when buying, that the latest purchase is the ultimate and final choice. Standard boats command standard prices, according to condition and inventory, but odd or idiosyncratic designs may be hard to dispose of. The market value of a boat depends greatly on the repute of the designer and moulder. A cheap hull will end up as a cheap boat no matter how conscientious the work and expensive the materials put into it.

Many moulders work under Lloyd's supervision and to their standards. The products of such firms can be relied on, although they will probably be a little dearer than others. However, this will be reflected in their resale value. If you wish, such firms can arrange for your hull to be moulded under Lloyd's oversight and given an individual moulding certificate. This is not prohibitively expensive and will raise the value of your boat, as well as having advantages for insurance purposes. Also, you will be assured that your hull is superlatively sound—a great comfort in times of stress afloat.

3

Buying

Before dipping into his pocket, an owner needs to have an unmistakable idea of how his boat will look when launched. All fittings and appointments should be congruous to their class and consonant with their surroundings. A miscellany of, say, bronze and glossy chrome or of stainless steel and galvanised fittings would look odd and might give rise to electrolytic troubles. Stainless steel rigging should not terminate, as I have often seen, in galvanised thimbles and copper swaging. Nor does galvanised wire match chainplates of stainless steel or bronze.

With metal fittings a choice is of cost and appearance. Stainless steel is often preferred for looks but there is nothing unserviceable or objectionable in fitting galvanised mild steel throughout. This may have to be of slightly heavier gauge than in stainless, but even so will cost considerably less. For a heavy-service cruiser it may even be better to standardise on galvanised mild steel which can be painted to match the general scheme and to give added protection against the elements. If you are designing and having fittings made up, you will find that really reliable stainless steel welding is very dear; any competent blacksmith can be trusted to turn out a sound job in mild steel which can then be commercially galvanised by the hot dip process. Blakes of Gosport, among others, specialise in this work; it is charged by weight. A faulty weld in stainless is not easily detectable to an untrained eye, and can let go without giving prior warning. Mild steel, on the other hand, stretches and it is usually easy to foresee trouble and do something about it.

Resinglass is so recent in the tradition of the sea that the eye has not yet become accustomed to the sight of large areas of what is regarded by some as 'bathtub' material. For those who will defend it there is sense in leaving the unashamed starkness of resinglass alone. It needs little maintenance and need not often be painted. Others may prefer to trim it to great or small degree with wood frames and surrounds. These are redolent of former boats and can be oiled, varnished or painted as seen fit, although there is not much point in painting such trim; better to paint the resinglass itself and save time and material. However, such matters must be settled before you can really visualise the final appearance of your vessel.

Once the overall picture is clear shopping becomes fairly simple. You should avoid the trap of buying bargains which will not fit into the scheme of things. A 'nearly' match for type or size could prove to be the proverbial ha'porth of tar. A boat is a durable thing. A change of mind or alterations made after completion could far outweigh the original petty saving on unsuitable items.

EMULATE THE SQUIRREL

As cash comes to hand it is wise to buy and store materials against inflation, future possible short supply and finicky difficulties of matching up. The costlier an item, the greater will be its proportionate rise in price, but keep an eye on trends. Plywood, for example, like most timber, suffers regular and sizeable increases in price. Woods like teak and iroko are becoming ever scarcer and dearer; if you get the opportunity to buy what you want, make every effort to get enough to complete the boat.

Although the cost of mast and spars, if of metal, may well be greater than the total cost of all timber used, their price is likely to rise only gradually and in smaller degree. Therefore give priority to timber purchase. Things like glass cloth and resin do not increase dramatically in price and are best bought *ad hoc* as work progresses. It might be wasteful to overestimate the amount required and buy in advance. There is also the question of storage and shelf life of resins; fresh supplies are always to be preferred.

Marine auctions are a fruitful source of chandlery so you should think of them as useful sources of supply. It is essential, of course, to discriminate in selecting goods and put a realistic limit on the worth of lots. Bargains do exist which represent real savings over similar items bought new from retail, or wholesale, sources. There is a lot of worthless junk to be picked over but I bought, at one local sale, new or fully serviceable items for £16 which would have cost more than five times that sum elsewhere. Where else, for instance, could I get a brand-new, double-acting Henderson bilgepump for £6.50? Or four new, eight-inch, Birmabright horn cleats for £1.25? And they match!

A glance at advertisements in the yachting press will reveal a great disparity in prices of materials and fittings between different suppliers; this is especially evident in the timber market. Even though having to pay carriage, it will often be profitable to order from quite far afield.

Before buying everyday items like cleats and blocks, consider that such items can be made from offcuts of

BUYING

timber which would otherwise go for scrap. Bits of hardboard and ply can often be ganged together with sticky tape to form parts of moulds for minor fitments—see pages 22 and 78.

TOOLS

A modest kit of wood- and metal-working tools is essential and should include:

Hand-, pad-, hack- and tenon saws.
Jack, smoothing and rabbeting planes.
Spokeshave; drawknife.
Wood chisels $\frac{1}{8}$-inch, $\frac{1}{4}$-inch, $\frac{1}{2}$-inch and 1-inch.
Hand- and breast-drills with wood and metal boring bits.
Counterbores with plug cutters to match.
Wood rasps, metal files and Surform tool with metal- and wood-working blades.
Screwdrivers to fit Nos. 6, 8, 10 and 12 screws.
Spanners and a Mole wrench.
Claw, peening and lightweight hammers.
Sanders.
Wood and metal vices.

This is a basic kit of hand tools, but of course today there are few who do not use the more efficient alternative of electric tools. Notwithstanding this there will be times when you have to resort to traditional tools.

When considering the purchase of electric tools, safety should figure large in your mind. Work on a boat is often in exposed and damp conditions, with an attendant risk of electric shock caused through leakage and bad insulation. Please, never use makeshift connections. It takes little longer to fit safe plugs and sockets which can be of the weatherproof variety. Also, doubly insulated drills are marketed which ensure that any leaking current cannot be transmitted to the body of the tool. This is the only safe type to use outdoors.

It is worth making a choice, if it can be afforded, in favour of individual tools rather than attachments for use with a general-purpose drill which, incidentally, will be put to heavy use when working with hardwoods. The more powerful the drill the less likelihood of burning out its windings. Self-powered saws are pricey, so that unless you are going to use a lot of unsized timber, the attachment may serve. On the other hand the self-powered jigsaw is relatively cheap, and invaluable in cutting curves in thick ply and framings. I mentioned the electronic speed controller before; this allows slow-speed work without loss of power and is preferable to mechanical reduction gears, whether built-in or attached to a drill.

I could go on at length about tools, but everyone will discover and solve his difficulties in his own way. One item I cannot pass without comment, as I found it almost invaluable in completing my own boat. This is the Black and Decker Workmate, a very versatile sort of vice-cum-sawbench-cum-staging. It is well-known and advertised, but I found every word to be true, which is not the case with many other widely publicised products.

No doubt most readers will have the nucleus of a tool kit already—what married man hasn't?—and it can be augmented as needs arise. Only *good* tools are worth buying. Cheap ones will wear out quickly, lose their efficiency, be inaccurately sized and cause bad workmanship and in every way be a source of trouble. A good workman has every right to criticise bad tools. Perfectly serviceable ones can be found at auctions for low prices, and any tools you have surplus on completion can be resold as auction lots. The price you get will depend on

BUYING

their brand name and the care put into their use and maintenance.

Heavy equipment like welding plant, belt sanders, disc grinders and so on can be hired from specialist firms. You should try to plan work to keep hiring time to a minimum and fit use of such equipment into the general scheme of progress. Vagaries of weather may mean that plans to use hired equipment have to be altered, so keep things flexible.

Simple tools for special, or one-off, purposes can be dreamed up and made by any average owner. Here are one or two:

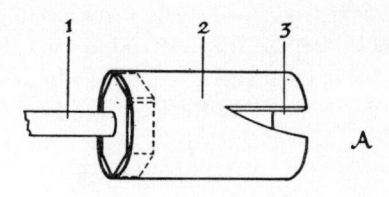

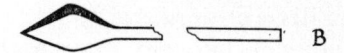

Homemade tools. A: A plug cutter. B: A flexible bit for use with an electric drill.

A: 1. Bolthead Araldited into body. Shank fits electric drill.
   2. Piece of gas pipe.
   3. Cuts made on both sides, tempered in gas flame and sharpened.
B: Bicycle spoke hammered and filed to shape, then tempered.

4

Planning

The sequence and progress of work will depend on many factors some of which, like illness or unforeseen shortage of supply, are not controllable. A reasonable allowance should be made for the unpredictable, but that apart, it should be possible to devise a schedule which will provide for work to be done regardless of weather even if your site happens to be in the open. And you will be very fortunate if it is not. Provided the hull has been made watertight at the beginning of fitting out, it should be possible to do quite a bit of interior fitting in all weathers. This gives an immediate indication of the priorities to be accorded to items of work.

The availability of cash figures largely if you are building in stages, and it is well to look as far as possible into the future. Work delayed for reasons of money shortage can, by incurring extra site rents and other periodic and inescapable charges, merely exacerbate the position.

Presumably any builder will have a pretty fair idea of how much time he will be able to give to his task, but it is possible seriously to underestimate the demands of the job. An awkward place into which to wriggle and put a nut to a bolt can cost you an unbelievable number of valuable minutes.

PLANNING

THE WORKING SCHEDULE

A list of work to be done should be prepared in as much detail as necessary, and by way of example I offer the list used in my own fitting-out. It is far from exhaustive, as I discovered, but gave me a firm basis to start with. It was hardly necessary to break it down into the number of nuts and bolts needing to be fitted, but even without such refinements it is dismayingly lengthy.

Against each item I put columns for probable duration of task (many were too small to bother about): date of starting; of offering up a component; of starting and completing alterations or modifications; of final fitting. This showed pretty well at a glance how the business was faring. What it did not tally was the worry, frustration and realisation that many things could have been tackled in a better way, less wasteful of time and energy. Everyone will have to allot his own estimates of time for tasks in choosing and buying, or designing, making and fitting every single item. And think of trips to chandlers, tool shops, auctions and so on from time to time.

SPREADING THE LOAD

Certain jobs can be put resolutely on one side for a rainy day. These include sailmaking, splicing rigging, making gimbals, mast fittings, crosstrees and a multitude of such small things which will not be needed until shortly before launching. Provided careful track is kept of progress of work on these jobs, you can lay them down at an instant to take advantage of fine conditions for bonding, painting and varnishing and so on. If you buy stuff well before you actually need to use it, fitting and construction can be flexibly planned. You should always have something lying around which can be picked up and worked on.

FITTING OUT A MOULDED HULL

Tasks are often interdependent and this means a little care in programming. You may make a galley unit at home and then take it out to the site for fitting behind a bulkhead. If this has been bonded-in and the unit cannot be manoeuvred into place you will find life a little dreary.

A garage, shed or room at home will have to be annexed in which you can work on, set down and pick up bits and pieces in various stages of construction, glueing, painting, drilling and so on. I found that I had to try to cultivate an orderly mind. Plenty of shelves and working space are extremely desirable. So is the acquiescence of one's wife to months of disorder, disarray and dust. This book is dedicated to a very patient wife indeed.

I had to be patient as well—no easy task—and keep remembering that despondency is to be avoided. Often it seemed to me that the task would never end, but it helped not to set any target dates. A missed date causes vexation and muffed workmanship as one tries to speed up work to meet it. If you can bear it, leave the date of launching happily nebulous until completion looms within credible sight.

A WORKING SCHEDULE

Exterior

Rubbing strakes   Handrail
Handgrabs
Cockpit:
    coamings   grid for bridge deck
    locker doors   grid for floor
    seat slats

Foredeck:
  cleats   samson post
  fairleads   chain pipe
  stemhead roller
Aft:
  cleats
  fairleads

Hatch surrounds   Window-frames
Hatch covers   Ventilators
Hatch runners   Rudder
Pulpit   Tiller
Stanchions   Self-steering gear
Guardrails   Trail boards
Windows   Navigation lights
Mounting for steering compass

(Omitted from my list were winches, fairleads and blocks for sheets, chain-plates, rigging and such other bermudan rig gear as I was fitting an unsupported junk mast.)

Interior

Ballast in keel
Glass over:
  keel
  all nuts from rubbing strakes, handrails, grabs and deckfittings
Cockpit drains   Chart table
Cabin sole   Domestic lights
Bulkheads   Wiring system and
Stringers   control box
Berths   Water stowage
Lockers   Piping

FITTING OUT A MOULDED HULL

Galley
Head
Shower
Decorative timber

Bilge pumps
Fire extinguisher
   mountings

Sailing gear

Mast
Truck fitting and
   other blocks
Boom
Yard
Sail

Battens
Halyards
Sheets
Downhaul and
   reefing gear
Sailcovers

Mechanical propulsion

Engine bearers
Shafting
Bearings and
   glands
Propeller
Controls
Instrument panel

Cooling system
Exhaust system
Tankage
Drip tray
Cover for engine

Other work

Mounting for
   navigational in-
   struments
Painting and decor-
   ating interior
Bunk cushions

Antifouling
Treatment of deck
   woodwork

5

The Working Site

The hull has been ordered, materials have been bought and stored, tools readied and a plan of campaign settled. Before the hull arrives you will have to find a working site on which to put it. If at all possible this should be near home, best of all close to the house in garden, barn, garage (although these are seldom high enough) or even in the driveway. If it has to stand in the open a temporary shed can be bought and later resold, or you can build a flimsier structure from timber or scaffolding covered over with tarpaulin or polythene sheeting of very stout gauge. At the worst a simple boat cover can be put over the hull when it is not being worked on. This will need to be securely lashed down and care taken to see that it cannot blow loose; the securing lacings must not be allowed to chafe the hull.

FACILITIES

You need reasonable access to a power supply and to creature comforts. If you are sited near home this may be easy to arrange and mean that tools and materials are near to hand. Additionally you may be able to press members of your family into helping with tasks requiring more than one pair of hands. If the work is distant you will be landed with the rent of the site and the costs of travelling to and fro. It is, unless you have unlimited time, virtually impossible to do without power tools so that the site must have electricity available to you. You will need water for hosing down the hull and for washing your hands. Toilet facilities should not be forgotten.

Storage of tools and material must be thought about. A kit of tools can be carried in the car but it is no simple matter to transport sheets of ply, large hardboard templates of curious cut and heavy containers of resin and paint to the site on every visit. You could buy a small trailer, or hire one, but it might also be possible to rent storage accommodation or build a small shed to sell off later. This brings the problem of protection of materials, also the hull itself, from the vandalism and theft so regrettably frequent today. Staging and ladders will be needed constantly during completion and their security has to be thought about. It is advisable to discuss the question of insurance against theft and damage in relation to the site concerned.

Looking to the end of the work, it will then be necessary to move your boat from the site to the place of launch. This will entail using a crane and truck, or at least a towing vehicle if the boat has been built on a trailer. Access at that time in the future must obviously be guaranteed: vacant ground could, in the interim, have been built on or otherwise obstructed. Unlikely, you may say, but such things have happened in the past and will again.

WORKING BASE

The hull will have to be set up securely on a firm, level base for both safety and ease of work. This can be a temporary construction of timber or scaffolding, or you might prefer to buy a trailer or cradle, depending on what type of boat you have chosen. I had a cradle built to my design by a local blacksmith for about £30. Such a tailor-made thing will very often be better suited to its purpose than a very much dearer proprietary article. If intending to tow, you will have to obtain, or make, a lighting board and fit the trailer and your car with eye and drawbar.

Hulls are often delivered sitting neatly in a returnable cradle on the back of a low-loader. Suppliers will sometimes sell these for little more than the cost of their timber content. Hull and cradle can then be lifted off complete from the delivery truck. If not, the hull will have to be taken out of its cradle and put on its base before the truck can depart. You will have to order a crane or see to other means of lifting, and arrival of lifting gear has to be timed to coincide with the arrival of the hull. Delivery times, by the way, can be wildly inaccurate either way. If the driver arrives early it is his own lookout, but a crane awaiting him will be charged for by the hour, expensively. It is worth giving quite a bit of thought to the organisation needed at this critical time.

For accurate and easy working, as will be revealed later, it is essential to set the hull up level in its bed. The receiving cradle or trailer should be sited on firm ground, which should not shift when the hull is first placed in position nor later as its weight gradually increases during building. You may have to dig the cradle in or shore it up beforehand; once the hull has been put on it jacks will be needed to alter its attitude. Ask the crane driver to lift and resettle the hull as often as necessary to get it quite level. It is probable that neither he nor the truck driver will have suitable slings aboard, and you would be wise to prepare a couple against the time, and make a pair of spreaders for use as shown.

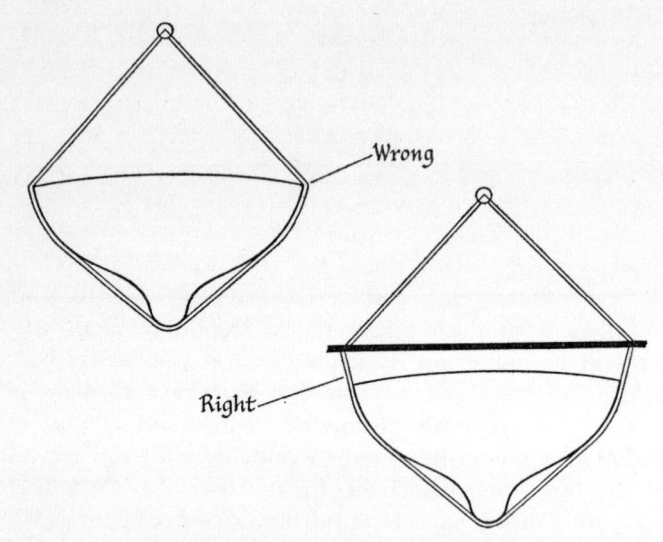

Use of spreaders when lifting hull.

These are essential to prevent the hull being pinched when slung; a very important point with a recent moulding. They should be of stout timber, say four by two inches, notched with a 'V' cut at each end and be slightly longer than the widest part of the hull at the points about which the slings will come.

Slings can be made up from synthetic rope with a breaking strain of about four times the delivered weight of the hull, or hull and cradle if both are to be lifted together. You can use nylon or Terylene rope, but if you have no further foreseen use for the slings, like converting them into springs or breast-ropes, it is cheaper to use polypropylene. Splice large, soft eyes into each end with not less than six tucks, dogged. Breaking strains are:

CIRC.	TERYLENE	POLYPROPYLENE	NYLON
in	cwt	cwt	cwt
$1\frac{1}{4}$	31	27	40
$1\frac{1}{2}$	45	38	59
2	80	65	104

INSPECTION

It is vital for you to inspect the hull as delivered before signing the discharge note. Once this has been done it could prove extremely difficult to have damage or defects rectified. In the extreme, although disappointing and unpalatable all round, it may prove best to refuse to take delivery and insist on the supply of a fully satisfactory replacement. Suppliers are not likely to despatch a dud hull, but damage of which they have no control or information can occur during loading or *en route*. If the hull has only minor blemishes or imperfections which are within your own capacity to put right, phone the suppliers there and then. Rather than chance having their delivery returned to them they might agree to some reduction in price. Ask them politely to confirm this in writing.

It is very easy to adopt a starry-eyed attitude when the long-awaited day dawns, but a hull costs much hard cash and you might as well get full value for it.

6

Preparatory Work

TOOLS

Satisfactory work can be done only with clean, sharp and fully prepared tools which are suitable for the job in hand. It makes considerable sense to allot time at the close of each working day to attend to them. Such action is an integral part of the work and not a tedious extra task. It is akin to a soldier cleaning his boots and rifle at the end of an exhausting day in the field—a discipline engendered by need and bred of routine. Such discipline helped me a lot during those dark days when joy and enthusiasm were replaced by careworn weariness.

Ancillary to this is the need to prepare for the morrow in other ways. If the site is distant, it will help you to make out a list of work to be done and the tools and materials that will be needed. If at all possible, these should be stacked or loaded the night before they will be used. A silly thing like forgetting the catalyst could make a day's glassing-in a write-off. A task put off for the next day can easily be omitted or overlooked in the urgency of departure because the sun is shining fair.

MARKING UP THE HULL

Datum lines will have to be marked inside a hull so that you have something to work to. Most of them are related in one way or another to the load waterline (L W L).

PREPARATORY WORK

Some hulls may have the L W L scribed or moulded in the gel-coat. My own has L W L and a boot-topping line scribed on, which is a bit of a nuisance because I don't care for boot-topping. If the L W L is absent you will have to mark it in either by scribing or painting with antifouling or boot-topping which will later be merged in. Such paint should not find its way on to the gel-coat of the topsides. A way of positioning the L W L is shown; the hull is, as we have agreed, level on its base, which permits the use of a spirit level (such as builders use) on staging and battens.

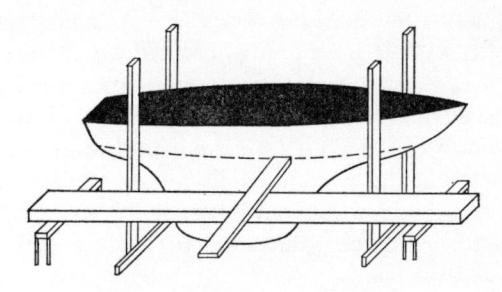

How to mark in a waterline.

Mark the ends of the L W L on stem and stern. These are simple to scale off from drawings, but if you don't have any you will know the working draft and can measure that distance up directly from the base of the keel, or bilge keels.

Put trestles so that staging will lie level with the two marks, and sufficiently below them to allow cross battens put on the staging to sit with their top edges level with the L W L marks fore and aft. Check, with a level, that the battens themselves are level. By moving them along the staging, continuously checking their levelness, you will follow the line of the L W L which can be marked

FITTING OUT A MOULDED HULL

intermittently and later joined up. Both sides of the hull will have to be marked. Take a bit of trouble over this fiddly job, because many of the interior measurements will have to be related to your marking.

TRANSFERRING THE LOAD WATERLINE

It is a simple matter to transfer your marking to the inside of the hull. Beg, borrow or buy a bar magnet and a strip of steel, or two bar magnets. Two people are needed, and this is a job to interest a younger son. Put him inside the boat armed with the steel strip (or second magnet) and a felt marking pen. Place your magnet along the L W L so that its top edge coincides with the marking. Tell the apprentice to move this strip around until it is attracted by the magnet, and then to draw a line along the top of it. With a system of Indian signs and knocks, he can be persuaded to follow the line as you pursue it outside with your magnet. A series of marks is left which can be joined up to indicate the L W L inside the hull.

When working in and about the interior of the hull, feet will scuff and erase the marking, so you may decide either to scribe or paint them on permanently.

PERPENDICULARS AND VERTICALS

The position of bulkheads and other thwartwise members will be perpendicular to the L W L and at right angles to the fore and aft centreline. The latter should be marked out accurately along the bilge. Fin keels usually mean a gaping hole right along the centreline; a length of whipping twine can be stretched taut over this and tacked in position with a drop of resin mix to join up with the marking fore and aft of it. Drawings show

PREPARATORY WORK

bulkhead positions, and also indicate which edge of the bulkhead has been used to give fore and aft measurements. Sometimes the same drawing will refer to either the forward or after edge of a bulkhead according to its station. This can give rise to a little confusion at times, and it is worth annotating drawings to make such points unmistakably clear. In a small hull, especially, an aggregation of half-inch errors can mean difficulty at some stage of fitting things other than bulkheads themselves.

When you mark the hull to show where the edge of a bulkhead will come, it helps to draw an arrow pointing in the direction of the other edge, thus:

Helpful marking at bulkhead locations.

This will act as a reminder to you to cover the arrow with the bulkhead when fitting it in, and so ensure that it is stationed correctly.

A line perpendicular to the L W L can be taken up the inside of the hull with a large set-square, T-square or builder's level, whichever is convenient to use.

FITTING OUT A MOULDED HULL

SYMMETRY

It is essential that all thwartwise members, including the very prominent bulkheads, are set at right angles to the fore and aft centreline as well as being vertical. Nothing looks worse than a feature which slopes or is askew, and any such will give rise to consequential fitting problems. Take check measurements constantly if you would avoid trouble. As a hull has virtually no straight lines this can be a bit tricky, so the centreline is used as a datum.

A length of inelastic line (an old log line is admirable for the job) can be held down on the C L by putting a chunk of lead or other weight over its end. This point should be six or eight feet away from the vertical lines you are going to check. Mark each opposing perpendicular at points equally distant from the L W L. Tauten the checking line and bring it to touch one of the points. When swung across to the opposite side of the hull it should touch the corresponding one.

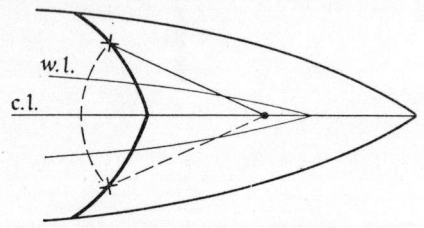

Line pegged to C L used for ensuring symmetry.

LOCATION OF COMPONENTS

Stations for components are shown on drawings as distances measured from stem, end of L W L fore and aft or other datum, which will be clear or stated. These are,

PREPARATORY WORK

it will be obvious, horizontal measurements in a straight line. They cannot therefore be measured along the curve of the hull or the rocker of the bilge. You can use one of two convenient methods of locating them and marking the hull. Stretch a taut line horizontally from fore to aft directly above the C L (it can be checked with a spirit level). You can tack it in place with resin, or fix it to small hooks glassed into position, which may be more convenient as you will often have occasion to remove and replace it. It depends on which part of the hull you are working on whether you take its forward end from high up in the bows or low enough to pass underneath the cockpit floor. Wherever it is, accurately determine its position and mark your drawing accordingly. Stations can then be scaled off along the line and a plumb-bob used to transfer them down on to the inscribed C L, from whence they can be taken out, in the usual fashion described, to the sides of the hull. They should then be checked for symmetry.

The other method is quicker but perhaps slightly less accurate. Fix a length of line to a point along the C L, well forward in the hull. This can then be used, as for symmetry checks, to mark spots on opposing sides by measuring off a length on the line scaled off from the drawing, which has been marked with the location of the fixed end. Obviously, once again the line must be held horizontally and this means that only one mark can be made on each side of the hull. These are extended into lines perpendicular to the L W L and you check them for symmetry using the same piece of line.

Of course, once a major bulkhead has been fitted, it becomes a simple matter to use this as a datum from which to measure forwards or aft, but keep in mind the fact that all measurements must be in the horizontal plane.

EXTERIOR MARKINGS

Deck fittings should be placed with an eye to symmetry both thwartwise and fore and aft. Cleats, deck-eyes, in fact all fittings paired port and starboard can be accurately stationed if you measure distances along and in from the side of the deck. It is particularly important to site chainplates correctly because rigging will not only look awful but will stress the mast unfairly if shrouds are not accurately opposed. A spot of paint, later to be covered by the base of a fitting, will not weather away and serves well to mark a position.

DRAINAGE

Once the hull is marked out you can start in earnest, but remember that condensation and incoming water will collect in the bilge. A drainhole should be made where the bottom of the sump will be (in the case of a fin-keeler) or at the lowest part of the bilge for other types. The diagram shows a simple way of fitting such a device. It need only be of small diameter, but as it will get choked by debris it should receive regular attention with a prodder.

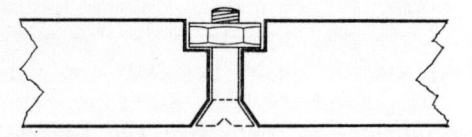

An unobtrusive drainhole in the bilge.

Special screwed plugs are available, but beside being expensive they are too large for the job and their use should be confined to dinghies.

7

Sub-Mouldings

Modules bought ready-made will need only to be bonded into place, but a lot of small fitments can be home-moulded either in place, or separately for bonding-in. The fabrication will give you practice in handling glass and resin, and I suggest that you should start with simple sub-mouldings before tackling any major items. Initial mistakes, such as we all inevitably make with unfamiliar materials, will then not prove unduly expensive of time and materials.

MOULDING A HATCH COVER

One of the plainest and simplest of mouldings is a hatch cover. This should be designed to fit snugly around the flanges of its opening in the deck. The length and breadth of the flanges should be measured, and checked against the diagonals. It is not at all unusual for apparently rectangular shapes to be a few degrees out of true due to inaccuracies arising during the moulding of hull and superstructure. Don't worry about this—it will not be obvious to the eye and no-one is going to measure around your boat to find out about it.

Accurate, doubly checked measurements must be taken before work is commenced. I don't propose to keep repeating this elementary advice, but you will find, as I did, that to overlook it at any time can cause all sorts of bother and vexation. Measurements are best drawn to scale on paper to confirm their validity. It is worth mentioning the fact that rulers and measuring tapes may not be absolutely accurate themselves—I own a beautiful, stainless steel metre rule which infallibly measures out 1,004mm for me. If you use the same instruments habitually, it does not matter if they are slightly inaccurate, as everything will be congruous.

To find the dimensions of the female mould needed for your cover, the clearance gap between cover and coaming will have to be added to your measurements; also the thickness of the cover walls. The results will show you the inside size of the mould; when cutting plywood or board to make it up, allow for their having to be overlapped at the edges.

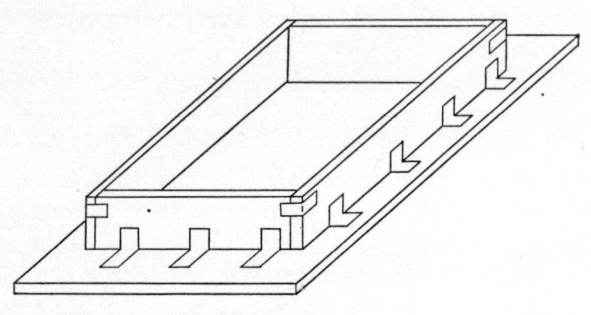

Mould for a hatchcover.

The mould is made, basically, from a flat sheet and four strips for the sides. These should be made $\frac{1}{4}$-inch or so deeper than the designed depth of the cover sides to allow for trimming. When learning to mould, I found it a little difficult to wet out fully right up and over the top of an edge, and it is important that the edges of a cover

SUB-MOULDINGS

are solid all through. Dry (i.e. not fully saturated) edges will damage easily and get to look ragged and scruffy. You can trim an edge easily enough with a sharp blade, if you catch it at the short interval between initial set, when it is rubbery, and going off hard. I find it difficult to judge this interval, and often let a moulding go hard and then trim it up with a hacksaw, Surform and wet and dry. As I mentioned before, slow-setting mixes leave a longer rubbery period.

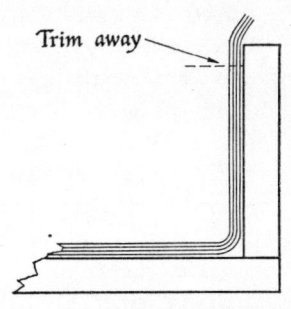

How to ensure complete wetting out of a vertical edge.

Hard edges should be avoided. They are unattractive and, more importantly, it is essential for you to work glass fully into corners and along edges otherwise these will contain only the brittle resin and be liable to chip. This means that your rudimentary mould has to be modified to allow for radiusing of edges and corners. As a gel-coat will faithfully reproduce every minute detail of its mould, time spent in perfecting the recipient surfaces will repay you in full.

Radiusing can be done by using any convenient substance to round out the coves, but you will have to fade it imperceptibly into the main body. Plaster or Polyfilla can be used to bulk out, but neither is easy to feather off on to wood or hardboard. Resinglass itself can be used, but is hard to shape accurately due to its fluidity, and when set is hard, to grind and polish. You can use Plasticene or putty, but take care not to press such soft stuff out of shape when putting on release agents or lay-up. I have, with reasonable success, used a mixture of half Polyfilla and half Cascamite glue, all mixed up to a putty with water. This is easy to sculpt, feathers well and sets like a rock without shrinking.

Take care not to leave crevices or interstices into which the gel-coat can be forced. This will cause lines of resin along the position of leak which may not release without bother and will, in any case, result in a defaced moulding when you grind them off later.

A resinglass mould is ideal in offering a perfectly smooth surface on which to lay-up, but for any one-off job the time and expense of the plug and mould system is not justified. However, instead of plywood or hardboard you can turn to resinglass sheets to form the basis of moulds having areas of flat surface. Ordinary sheets of

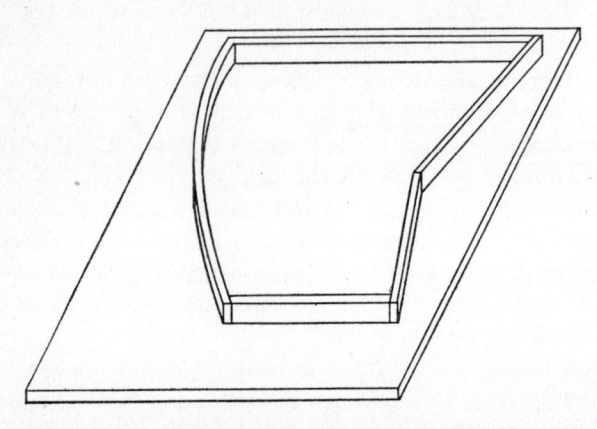

Mould for a resinglass flat laid on a sheet of window glass. Can be stuck in place with Sellotape.

SUB-MOULDINGS

glass, which need no release agent, can be used to lay-up resinglass flats which can be made to accurate size by using simple shuttering.

Alternatively you can lay-up a fairly large sheet and then just cut it into strips and shapes with a jigsaw.

In using such pieces of flat, you should be careful to ensure that they are not irretrievably damaged by getting stuck together or having their surfaces defaced with raw mix, otherwise they may not be reusable. Gapfilling and radiusing should be done with materials which will not damage the resinglass used for moulding.

One-dimensional curves can be given to a mould by either cold-moulding or springing plywood into shape. Thin resinglass flats can be sprung into fair curves and then immobilised by adding enough extra laminations at the back, but they may then be unfit for further use.

ROUTINES

Once the mould has been made and faired up you will have to prepare it for mould release. Plywood and hardboard, or any other porous surface, must be sealed with Ronseal, Bourne Seal, polyurethane paint or varnish or something of the sort which will give a smooth finish. On top of this, or straight on to non-porous surfaces like resinglass, you need to apply a couple of coats of the first release agent, a kind of wax; polish each application to a smooth finish before proceeding. By eyeing the mould surfaces from different angles you can spot if any small areas have been left untreated; lay-up may adhere to them. The whole of the mould surface then has to be covered with the second release agent, normally a polyvinyl alcohol liquid which rapidly dries to a hard finish. Unless care is taken, this coating is liable to retain brushmarks, later to show up in the moulding, and you may prefer to sponge it on. If you are using a resinglass mould, the customary release agents can be forgotten, and the mould covered with Slipwax. This is applied as directed and polished smooth. The technique is best suited for repetition moulding and is unsuitable for other than pure resinglass moulds, so you may not encounter it.

The next thing is to put the gel-coat directly and evenly on the prepared mould at a rate of not less than $1\frac{1}{4}$lb per square yard of mould surface. It should be allowed plenty of time to set fully; thereafter the exposed surface will remain a little tacky, so do not be deceived by this.

While it is setting you can be about cutting out CSM into pieces convenient to lay in to the mould. These need not be cut accurately, and there is even a slight advantage in leaving edges ragged and uneven so that pieces will merge together imperceptibly. Avoid any butt joints, and cut pieces large enough to overlap by about an inch. If you take care to cut pieces so that overlaps can be staggered, bulges will be avoided.

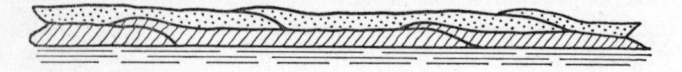

Staggering overlaps.

It does not matter if you find that coves and corners get thicker than walls and top; this is mechanically advantageous, but take care not to close clearance gaps. Cut three pieces of each shape, and the resulting $4\frac{1}{2}$oz lay-up will be amply strong for a hatch cover. Finally offer up the pieces for fit and match.

Next, measure out lay-up resin in the quantities needed. Full wet-out calls for 2lb of resin to 1 square yard of $1\frac{1}{2}$oz CSM, i.e. 6lb per square yard of hatch cover. It

SUB-MOULDINGS

does little harm if you use a bit more than the specified minimum, but a great deal too much will mean resin-rich, and therefore brittle, lamination. If you have resin left over, this will mean inadequate saturation and a suspect lamination. It is not wise to mix too much resin and catalyst at any one time, and I like to divide my resin into 1lb batches, each in its own marmalade jar. If you use a dropper, or syringe, it is simple to add the right amount of catalyst whenever you need a new batch of mix. Paste is best squeezed out on to the stirring rod—excess can be scraped away.

Everything is now ready for the off, and only minor things are left to be done. Get the container of brush-cleaner readied, coat the hands with barrier cream, take a deep breath and gird the loins.

LAYING-UP

On top of the gel-coat you now should spread a thick layer of mix. It is a common mistake to spread it thin, like paint, but avoid doing this because you are using a building material, not a decoration. Put pieces of mat into position and, after allowing a short time for the underlying resin to start seeping through, stipple away; add more resin evenly until the full amount has been used up. More stippling, rolling, stippling and rolling has to follow until the resin has been saturated. It is then wrong to put on the next layer of mat, even though the treated surface is obviously glistening wet with resin, because the full amount of mix appropriate to the second layer will all have to be worked down through the top. Instead, add another thick spread of resin before laying down the mat.

As you reach the end of a container, put the working tool into the cleaner and use a fresh one.

FITTING OUT A MOULDED HULL

During stippling the mat will slide about quite a bit and needs to be pushed around gently until it lies evenly over the mould. It will tend to lift away from edges and corners into a curve of greater radius than wanted. Press it firmly back and do not worry if this causes overlaps to be rather scanty in spots; you can adjust the situation with the next layer. Be careful to reinforce corners properly—get the mat well down into them. The mat and resin should be carried up and over the edge of the mould sides. Once the whole of your material has been used up the moulding should be left for the resin to go off hard, unless you propose to trim during the rubbery stage, of course.

RELEASE

Strips of hardboard inserted between mould and moulding will allow the vacuum to break and often the moulding, if of simple shape like this cover, will loosen easily. If not, as is frequently the case, thump the mould with your fist, or a mallet. If this is a one-off job you might find it simpler to dismantle the mould if the moulding is tedious to release. However, if you have been thorough about putting on the correct release media, a moulding will usually come away without excessive bother. Do not use wood or metal to prise with as they will damage the gel-coat.

Many items can be moulded directly to the hull, e.g. tanks and lockers. The hull itself acts as one side of the fitment, and the remaining open-sided mould required can be fixed to it with little trouble.

TO MOULD IN A WATER TANK

Where such a fitment is concerned it is often easier to

make the mould of stout plywood which is left as an integral part of the structure instead of being used as shuttering and removed afterwards. In this case you need only bond the edges to the hull with the requisite number of strips of mat, or tape, and coat the interior with tissue and finishing resin to provide waterproofing and the smooth surface which will help to prevent fungal accretion. The exterior can either be skinned with mat or just painted over.

If, when moulding on direct, you want to see a presentable surface exposed, you can use resinglass flats with their gel-coat showing. The main bonding can be done on the inside, and careful work with tissue and finishing resin will disguise the exterior joints. Alternatively you can use a wood trim or cornice with good decorative effect.

8

Starting To Fit Out

In this and following chapters I suggest a logical sequence of work—not the only one, by any means—and try to bring out salient points not already covered.

You will first want to get your hull watertight so that work can go on in all sorts of weather. Although I deal with one area of a work at a time, for the sake of tidiness, there will inevitably be some dodging about between jobs. The progress chart I used was a great help in keeping track of events. I had to take care to time the installation of bulkheads and other large fitments so that I was not hindered for lack of access and room to operate. One cannot be specific about a matter which will obviously depend on the individual accommodation being worked on.

RUBBING STRAKES

Fit these first to stiffen up the hull. Up to a thickness of half an inch or so they can be pulled into place without steaming although the timber creaks alarmingly. These strakes should be knot-free as far as possible, or they may crack under such unprofessional treatment. If you prefer to have them thicker than this, they can be steamed, which is a dreary process, or more simply laminated from lengths of half-inch timber glued together during the process of fitting them. In this case the bolts through the underneath lamination should be staggered and so spaced as to fit in with those of succeeding ones. Bolts through underlying laminations should be with countersunk heads let in level with the surface of the wood. Counterbore for the bolts entering the outer face of the strake so that matching plugs can be glued in and cleaned off. I learned a little late that plugs with the grain not carefully aligned with that of the strakes are incredibly prominent to the eye.

Bolts should be spaced not more than one foot apart for a single strake and eighteen inches (staggered) for laminated items. Bolts at stem and stern should be fairly close to the ends of the strake—say about an inch—and backed up with another one set about two inches along. This is for two reasons. When bending a strake around the sheer much leverage is applied to the end bolts. Later on, when afloat, it is the ends of strakes which are likely to pull away under impact and added security is a wise precaution during the building time.

In measuring to drill strakes, make sure that holes drilled for bolts will not coincide with any rivets which may have been used for purposes of bonding hull to superstructure. Take measurements of rivet positions around the curve of the topsides and lay off boltholes accordingly. It is nice to have your plugs at regular intervals, but the odd one which is offset an inch or so will not be very noticeable.

Leave the strakes some inches—perhaps as much as a foot—aft of the transom to allow for manipulation and, after, trimming off or fitting a transom strake. Any chamfering done in the latter case should be on the side strakes to prevent the transom rubber being displaced by a glancing blow along the side of the boat.

FITTING OUT A MOULDED HULL

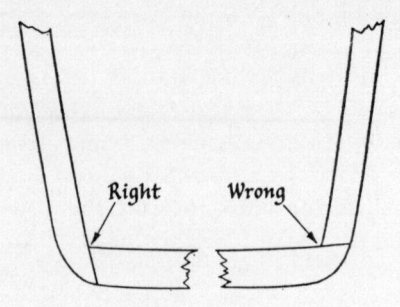

How to chamfer rubbing strakes.

Lash, or get some hands to hold, the strakes in place so that you can drill through the hull for the insertion of three or four bolts forward. Put these in and bolt them up securely, a job which requires someone inside and someone outside the hull. By means of a Spanish windlass the after ends of the strakes can be closed up against the transom so that the whole length of the strakes lies along the hull. As you then continue to drill from fore to aft, you will need to lift and lower the assembly so that the holes follow the line of the sheer. It is best to deal with one strake at a time. Each bolt should be inserted and tightened as you work aft, and it will be necessary to strain the strake to follow the sheer.

There are many ways in which hulls are joined to their superstructures, and many of them leave quite a gap on the outside of the boat. The strake must be laid to cover this. Also, throughbolts should take in both parts of the assembly, so that you may find yourself drilling through two or three thicknesses of resinglass. At times the join will be stiffened up inside with a stringer of bonded-in timber. If the strake bolts can be put through this it will ensure a very resistant deckside. It is advisable to seal any holes through such timber with resin mix to prevent water entering the stringer and perhaps giving rise to rot later.

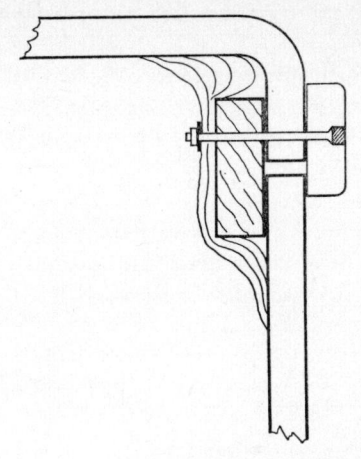

Using strake to cover gap between hull and superstructure.

The simple addition of the rubbing strakes turns a hull from looking like a moulding to looking like a boat, even though a bit embryonic, at last. You will be well pleased and breathe a sigh of satisfaction.

This is, regrettably, premature, because you have to take the strakes off again. Then you have to put them back on again! This time with a thick coat of mastic between strakes and hull. Unless bonded on, all wooden and other fittings will, you will discover as incredibly as I did myself, have to have mastic put between them and the resinglass of the hull. It keeps water from entering bolt and screw holes, and thus stops woodrot in the fittings and leaks in the hull. Additionally, it acts as a buffer between stressed fittings, mainly metal ones, and the vulnerable surface of deck and hull.

If you are laminating strakes, the lowest strip has to be masticced and bolted home. Its upper surface should be coated with glue and the bottom of the next one to be put into place liberally spread with hardener. If the weather is warm the top strips will have to be bolted on quickly before the shuffling time of the glue expires.

Shanks of bolts protruding inside the hull can be snapped off short with a Mole wrench or sawn down level with their nuts. You can then cover them with a strip of tape of 1oz quality or 1½oz C S M. I prefer to put separate patches over each nut. It takes a little longer but results in less defacement if a bolt has to be replaced later, as only a small bit of resinglass has to be chiselled away.

HANDRAILS AND GRABS

These items should similarly be throughbolted, taken off, set on mastic, and their nuts patched. I found them invaluable when working around a deck some height off the ground because my sense of balance is not outstandingly good. Especially the morning after.... It would perhaps be even more helpful, and secure, to fit pulpit and lifelines next, but I started on the ballasting of the keel first.

KEELS

The addition of ballast dramatically lowers the C G of the structure. Once in place, you can put a board over the sump, perhaps temporarily tacked into place with a couple of strips of mat, so that working around down below does not carry the risk of stumbling into a hole.

If the keel is a casting to be bolted on, you will have to jack up the hull to get it into place. A thick layer of

STARTING TO FIT OUT

mastic, a thin neoprene gasket, or both should be set on the top of the keel before the hull is lowered on to it. Tighten keelbolt nuts on thick, large washers; these can then be patched with resinglass. The bolts will usually have been set into the casting, but if this is merely bored out to take long bolts from the base of the keel their heads should similarly be glassed over with wet mat or resin putty. A casting can be fully sheathed with 2oz C S M covered with tissue and finishing resin, but at the least a strip of mat or tape should be bonded over the junction of hull and keel to stop bolts corroding. If you want to get at the bottom of the keel to sheath it, make sure that your jacks are man enough to take the increased weight.

Castings of lead or iron to be inserted in hollow keels will usually be of such a size that they will go down through the main hatchway. They may, however, be so heavy, even sectionalised, as to call for the use of lifting gear. Before they are lowered into the keel, its inside should be coated with enough wet mat to allow the castings to settle firmly into place. Once they are home, you should cover their top with laminations taken up six or so inches along the sidewalls of the keel. These laminations should be about as thick as the wall of the keel at the top edge of the ballast—typically 10oz or 12oz of C S M. It is imperative that no ballast should ever get loose. If capsized you will find it bad enough without being showered with lumps of lead or iron castings.

The more common kind of filling is lead ingots or iron pigs (or is it lead pigs and iron ingots?). These can weigh variously from 20 to 50lbs each and can be manhandled into position. They can be immobilised with a pour of resin or concrete; the latter is much cheaper and equally effective. Concrete bonds quite well to cured resinglass, but you will make a better job if you trouble to coat the inside of the keel with resin mix immediately before pouring the aggregate around the metal. Instead of ingots, which are not cheap, into your concrete can go such things as boiler punchings, scrap metal, superannuated mangles and so on. It does not matter a bit if they are rusty, as no further oxidation will take place once they have been immured in the aggregate. My keel is a very interesting object indeed, and contains lead shot, lead pigs, iron firebars, and a great quantity of used shotblasting pellets. I used these pellets, which looked and felt like oversized iron filings, instead of sand in making some of my concrete pour. It weighed an incredible amount—I could only lift half a bucketful at a time —and went off like a rock. The effect of all this jugglery on my steering compass is awaited with interest, not to say trepidation.

You are advised to divide your keel cavity into sections by bonding in webs of ply or mild steel before fitting the ballast. Bond them in with 3oz of C S M and they will strengthen and stiffen the keel.

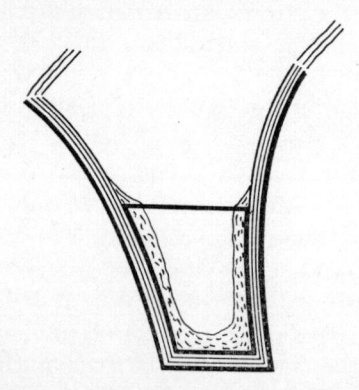

Plywood or metal web bonded into keel cavity.

It is sometimes held that these webs should be perforated so that the ballast homogenises, but provided the material is all densely packed this seems debatable.

COCKPIT DRAINS

A self-draining, and thus watertight, cockpit will cause rainwater to run down through the drainholes into the bilge. Once you have put your ballast in, maybe before would be more sensible, cockpit drains should be fitted to conduct water outside the hull. It is best to end the drainpipes at seacocks; as the outlets are below waterlevel a means of shutting off accidental inflow is advisable. Skin fittings of any type can be replaced or removed for maintenance providing they are not bonded in, which makes the operation very difficult. Bed them on mastic as you would in any other sort of hull. Any risk of crushing the resinglass of the hull can be averted if you set them on a thin neoprene gasket with mastic on both sides. Be careful not to overtighten skin fittings with large, gas-type nuts. Seacocks should be mounted on wooden pads of about an inch thick which can themselves be bonded on. If you put a seacock directly next to the hull, even if correctly bedded, there is risk of damage if it gets corroded and stiff to open and shut.

HATCH COVERS

These can be moulded or fabricated and, together with washboards or doors, will finish off the task of weatherproofing the hull. Fore and aft hatch covers should close down on a rubber or neoprene gasket as a weather seal. A forehatch should be hinged on the forward side. Hinges are often badly stressed, and the wall of a hatch and also of the corresponding flange or coaming should be thickened up to take them. This can be tricky, because if the hatch is closely fitted, the thickening will have to be on the outside of the hatch wall and the inside of the flange. Hatches bought often have thickened areas for the hinges; more often they do not and you will have to make as neat a job as possible. Small blocks of suitable shape can be moulded and coloured to match the hatch cover and, after abrading the affected areas, stuck on with wet mix; I prefer to use epoxy resin for these small but essential jobs. If you mould your own covers, remember to design them to accept hinges.

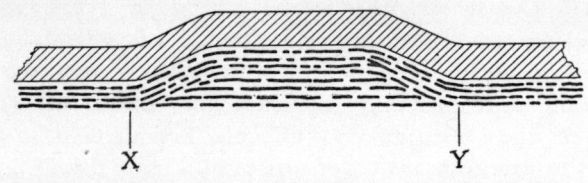

Section of mould and laminations in way of hinge fitting on hatch wall. Do not permit laminations to follow line of mould between X and Y, or the required thickening will be reduced.

Loose hatch covers should have no place on a seagoing craft, as there is always the danger of their coming adrift in hard conditions. However, if you choose to have them, perhaps for a river cruiser, they should be fitted with means of securing against weather and intruders. A strongback, as illustrated, is as good a means as any of pulling a cover down on a seal.

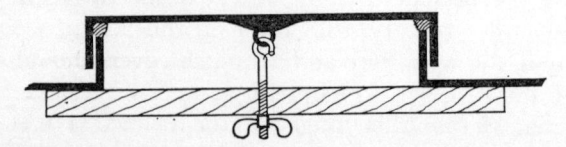

Strongback to retain loose hatch. Bolt taken through stiff beam of hardwood.

The cover will need to be thickened with moulding or a wooden pad bonded in.

Some hatch covers, like those on small foredecks, are trodden on, sat on and jumped on throughout their life and it is well, if you will take the trouble, to reinforce them against too much flexing which will eventually cause the gel-coat to crack or craze. Fibreglass is naturally flexible to a surprising degree, but can be stiffened up very easily. There are many ways of doing this, but the easiest effective method is to add a channel section which acts as a beam.

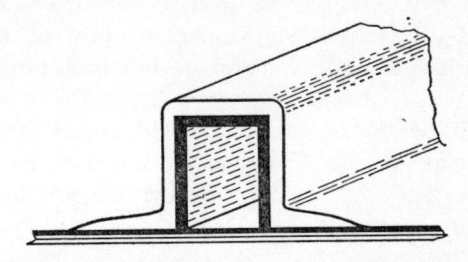

Channel section moulded over cardboard or plywood former.

Such devices can be used to add stiffness to any flattish areas such as cockpit soles, coachroofs and so on. The glass is moulded over a simple former of cardboard which can remain inside the channel after moulding.

Runners for sliding hatches must be throughbolted, and I think it best to stick them on with epoxy resin as well. This can be interposed between any well-fitting runner and the coachroof without the addition of glass, as it is being used as an adhesive and not a constructional material. Araldite glue would serve but be more expensive for the job. Clamp the runner in position, run round it with a pencil and stick masking tape outside the lines. If you pull this away before the epoxy resin has gone off the edges will be clean and professional looking. If makes no difference if the runners are of wood or of metal—epoxy resin will stick both materials adequately. I said a well-fitting runner, because wooden runners involve compound curves and sometimes will not lie absolutely true to the line of the coachroof. Mine didn't. In such a case you may find it better to use mat in conjunction with the resin to fill in any small gaps. This goes virtually unnoticed.

Boats flex quite a bit when sailing, so the opening for a sliding hatch should be stoutly reinforced along its flanges. It is possible, of course, to buy or mould a strengthening module to bond in, but moulding is such a complex job that it is always easier and cheaper to frame the inside of the opening with wood. The frame can be made to match a wooden hatch cover, if your tastes are that way. The inside corners are best dovetailed and glued.

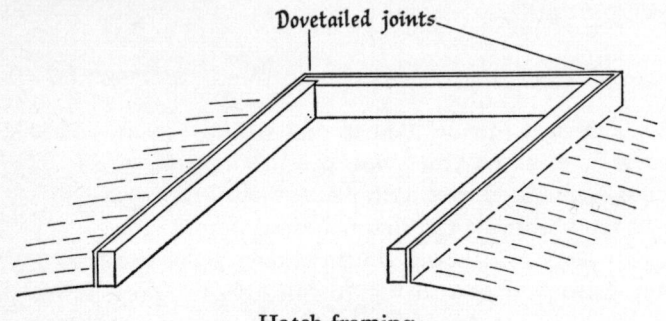

Hatch framing.

As this entry point is in constant use and also very prominent to the eye you should take a pride in its appearance, as well as its strength.

Some coachroofs have sections raised on which are

meant to fit lengths of tracking over which the hatch will slide. It is as well to check that these sections are thick enough to take the bolts holding down the track without reinforcement. When being moulded they are in fact long cavities to be filled in, and tend to be resin-rich and scant of glass as was mentioned in connection with moulding a hatch cover.

LEAKS

Every hole made in a hull is a point of potential leak and should be treated against the ingress of water as soon as possible after being made. All nuts should be glassed over, as explained. Every effort should be made to minimise the number of throughbolts fitted. As an example of what can be thought about, my own boat has a foredeck fitting made as a unit. This is a five-foot length of eight-inch by two-inch timber on which I bolted separately three fairleads, three large cleats and an anchor bollard—a total of 28 holding nuts and bolts. The assembly fits to the foredeck with four $\frac{1}{2}$-inch and four $\frac{1}{4}$-inch bolts, meaning only eight holes in the hull as against 28. Stress against the larger-diameter bolts means less shearing effect on the resinglass of the deck. Should a fitting let go the damage will be confined to the timber of the unit, which can be removed complete for rectification, and not to the hull. By siting cleats in pairs and treating other fittings similarly I managed to keep the piercing of the hull down as far as practicable. Even so, there is a very large number of unavoidable holes throughout.

9

On Deck

VENTILATION

A closed hull needs ventilation to reduce condensation. Dorade boxes are neat and effective. I moulded a pair using a polythene foodbox as a mould for $4\frac{1}{2}$oz C S M on top of gel coloured to match the topsides. The drainholes were cut out with a jigsaw, as were the openings to take the cowls.

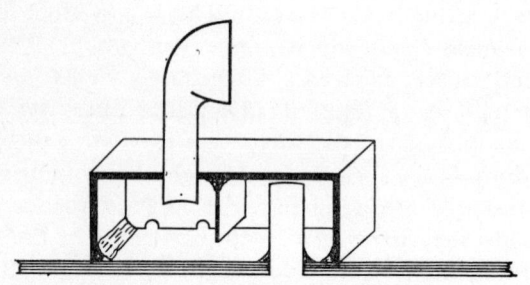

Section through Dorade-type ventilator.

If these boxes are to be set on the coachroof, it is a simple matter to fix a navigation light to their outboard side.

The inlet pipe can be of any sort of tube—aluminium is easy to work—or you can mould a tube around a cardboard former round which polythene sheet has been rolled. Use two layers of $1\frac{1}{2}$oz mat and cut out the former when the resin has set. Cut out holes of the right size to accept the tubes in the top of the deck, or wherever; these need not be accurately sized, for the tubes can be bonded into them quite roughly—they will be invisible. Bond baffles of $\frac{1}{4}$-inch ply into the box and make inside flanges of either ply or laminate, as shown. The cowls are screwed home. Position the completed assembly over the inlet tubes, pencil around and mask with tape. Bond the flanges on the hull with either polyester or epoxy resin. I used the latter for added security, as ventilators get the occasional thump from sheets and feet.

If you like the flat sort of ventilator—Tannoy or similar—as being less obtrusive, you will have to cut fairly large holes in the hull, which can be potential sources of weakness and leaks. If doing this, you should thicken up the edges of the holes and seal raw resinglass with a lick of mix. It is often preferable to bond on a wooden frame to the inside of the coachroof to match interior trim.

GROUND TACKLE FITTINGS

Every item connected with ground tackle needs to be extremely stout. Bollards, cleats, winch, fairleads, chainpipe, stem roller and so on should be firmly bolted through stout backing pads with mansize bolts and nuts. Anchor chocks are often just bonded on, but this is not secure enough and they should be throughbolted, perhaps not quite so massively.

TOERAILS

It is sometimes possible to combine rubbing strake and toerail in one piece, or two sections glued together. If not, you should throughbolt wooden toerails. You can mould lengths easily by halving an aluminium tube longitudinally and treating it for release in the normal way. Lengths need not be extreme, and spaces left between them will act as scupperage. Mould them of $4\frac{1}{2}$oz C S M and add a connecting tongue every foot or so.

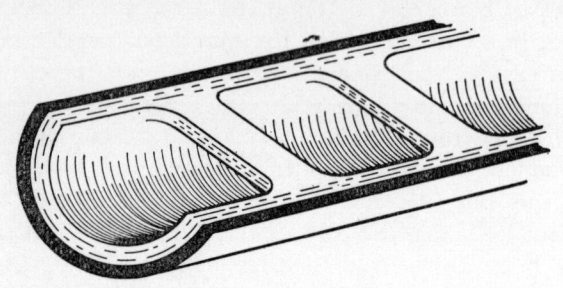

Toerail moulded inside length of metal tubing. Note tongues for bonding on to hull.

The rails should be epoxied on to the deck along their edges and tongues. Moulded toerails are of dubious advantage because they get scruffy with use; lines are constantly passed over them, which chafe, and they get kicked as a matter of routine. Unless buying a hull with such items moulded in I would always go for wood, and in any case add wooden caps to moulded rails to take the rub.

COAMINGS

Lines are also led over cockpit coamings, and in addition gear is rested on them and they are subjected to very hard wear. They should be protected with wooden cappings. If the coamings are flattish on top this is no problem, but those with rounded tops are hard to deal with. Cappings should be routed out to fit and stuck down with epoxy resin. You will find that it is not really practicable to modify the top of a round coaming to a flat section.

A buildup of resin looks unsightly and I doubt if it would be of much use to cut the tops away in any attempt to flatten them; it might dangerously weaken the structure.

On my own boat the handrails along the coachroof are set on two-inch high pillars about a foot apart into which bolts are set. For the cockpit I made similarly-shaped pillars of wood and fitted the cappings along the top of these. They not only prevent scuffing but serve us useful grabs. They are also very ornamental.

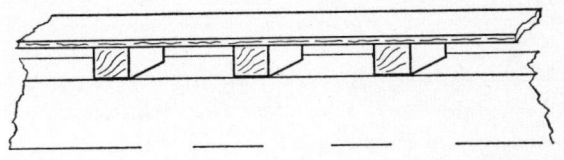

Handrail cum capping set around cockpit coamings.

WINCHES

Sheet winches must be stoutly anchored. Some small craft are not fitted with moulded winch pillars and you may decide to add them. Moulding is not worth while and you should find it better to make them of solid hardwood, bolted and glued into place. I have known them bolted on over mastic, but this seems to me to be wrong

Space for water to run between winch pillar and cockpit coaming.

FITTING OUT A MOULDED HULL

for such a stressed item. If you fit them, leave a space for water to run between pillar and cockpit coaming or water running along the weather deck will hit the obstruction and spill into the cockpit.

Winches fitted to moulded pillars should have thick pads fitted below deck for the bolts to go through. No matter how solid the top of the pillar is, unsupported bolts will work in the resinglass and create leaks. Plenty of mastic is needed under both winch and pad.

CHAINPLATES

Stresses transmitted through standing rigging are high, and chainplates need very careful attachment. Types of these fittings are shown, each one of which will cause a shear stress to be applied to a small area of the side of the hull.

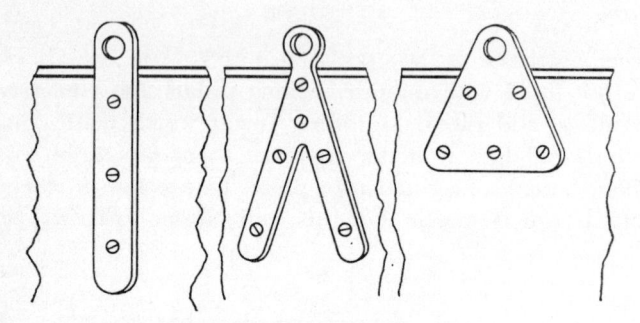

Type of 'bolt-on' chainplates.

Unless these have been moulded in at the factory, when they will be virtually integral with the structure, you would be better advised to fit types which pull up against the underside of the deck, like this:

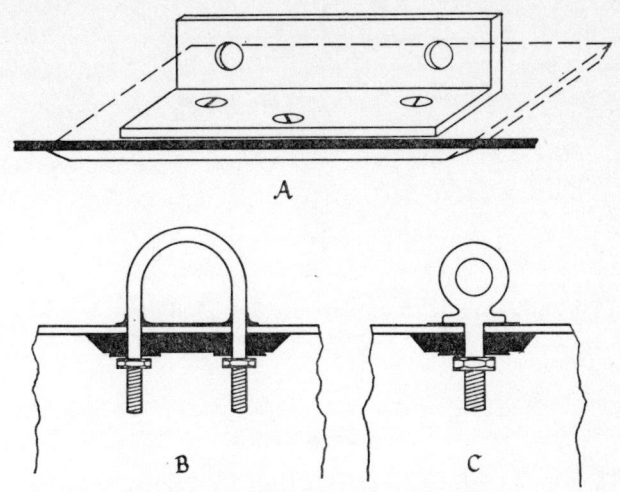

Recommended types of chainplate.

These convert rigging stresses into a compression load which can be distributed over a fairly large area; resin-glass takes more kindly to this than to shearing stresses. Also, replacement is easy in case of damage in contrast to moulded-in items which would mean extensive repair. The U-bolt model is neat, especially in stainless steel, but the good old-fashioned eyebolt of galvanised iron is hardy and trusty. Whatever you use will have to be backed by a large pad.

Forestays are usually taken to the stemhead fitting which is customarily strengthened against ground tackle loads. If taken to a point aft of the stem they can lead to U-bolt or eyebolt. The same considerations apply to attachment points for backstays. A chainplate sited close to a bulkhead can be tied in to by using a metal angle-plate. This spreads the stresses most satisfyingly; if a continuous chainplate is fitted (type A) this can often be linked to a bulkhead as shown.

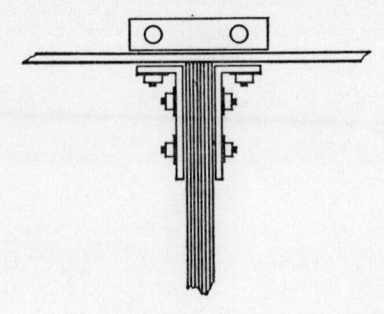

Continuous type chainplate tied in to bulkhead below with angle straps. Affords excellent way of spreading stresses.

MAST STEPS

Deck-stepped masts usually fit into a socket or tabernacle on the coachroof, bolted through a thickened area of the moulding. In some cases a depression is provided in which the heel of the mast rests. In many boats a supporting mast pillar of metal extends from the underside of the deckhead down to the keel. If this pillar fits into a socketed plate at its top end, the bolts holding tabernacle or step usually pass through both fitments. I would not bond either step or socket in place because of possible future need for removal or replacement.

THROUGH MASTS

Where one enters the deck it will need a mast-coat and a flange may have been moulded to fit it around. If this is absent, it will be simple for you to mould one on. It should be stout enough not to distort when mast wedges are driven in around it—say about $3/8$-inch thick. Wedges are best of softwood or hard rubber which will not damage the flange.

TRACKING

All tracking must be bolted on without bonding as you may well need to move or replace it. Put pads beneath to spread the upwards component of sheet loads.

PULPITS AND STANCHIONS

There are many ways of fixing stanchions to decks and a lot of them are really not very satisfactory. The most common method is to socket a stanchion in a plate which is bolted to deck, toerail or both. The plates are designed to fit close to the toerail so that they will not further encumber sidedecks already too narrow. The leverage exerted by loads applied to the guardrails or tops of stanchions ultimately loosen the holding bolts and leaks are common beneath them. The deeper you can bury the root of the stanchion the more security and less likelihood of leaks you will have. A good way of mounting stanchions is to mould sockets below deck into which they can be dropped.

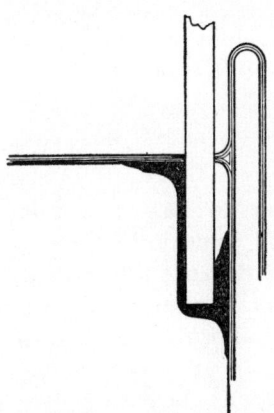

Socket below deck to accept stanchion.

This sort of socket will always be filled with water, either fresh or salt, so you will have to attend to the waterproofing of any galvanised stanchions. Although it seems a good idea, on first thoughts, to bore little outlet holes through the hull, this would cause lines of discolouration to form. Even resinglass will eventually show signs of age.

If your boat has a deep toerail moulded in, the stanchions can be set in sockets moulded against it.

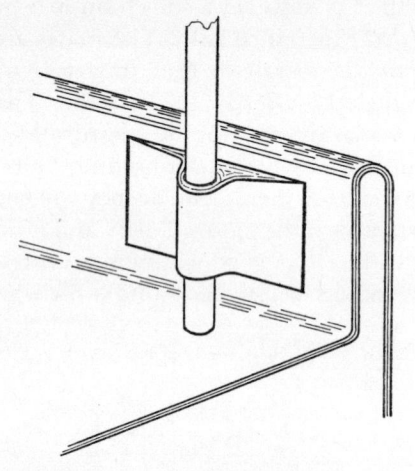

Stanchion socket moulded on toerail.

It is only sensible to pass a stainless steel split-pin through socket and stanchion for retention. Also to end the socket a small distance above the deck to allow for drainage. In this case water will just drain along the scuppers and should leave no more marks than other drainage of the sort. It is best to mould such sockets around the stanchions themselves, which are protected from adhering to the resinglass by coating them with grease or wrapping them with polythene. I like the polythene wrap as its thickness makes it easy to remove and then replace the stanchions.

Pushpits and pulpits can satisfactorily be bolted in position—they do not flex like stanchions. They are usually bought with feet welded on, but it is simpler to make your own, without feet, and fit the ends of the legs in sockets like those for the stanchions. As they will not be held down by the guardrails it is essential to pin them through for security.

Wiring for navigation lights can be wangled out through a pulpit leg and be taken invisibly down through the deck. If you use moulded sockets, remember that the wire should be led through a rubber grommet set into the resinglass to stop leaks.

BOWSPRITS AND BUMPKINS

Some craft are fitted with these and their means of attachment need thinking about. Should a bobstay let go the pull upwards on a bowsprit could be large, and things have to be arranged so that the sprit breaks before the deck can lift and be affected structurally. The area round the forward strap, which is where the pull will be exerted, must be massively strengthened with timber and laminations. This will also serve to resist sideways forces. You will need to ensure that the sprit is so held as to resist the backward pull of luffwire and bobstay, but let go and slide aft if you ram anything. How this is done can only depend on the individual boat, but it is a cardinal principle that any fitting in resinglass, if vulnerable to damage, must break before the structure of the boat is threatened.

RUDDER FITTINGS

These are vital to the safety of a boat and their integrity must be unquestionable. Keel-hung rudders present few problems; the stock leads down through a metal or moulded tube which also acts as a strength member. A rudder hung on the transom is susceptible to damage of various sorts and you must pay attention to its fittings.

Under no circumstances should pintles or shafts be affixed to an unreinforced transom. Stresses can be very severe, so all bolts should be taken through a stout timber laminated on the inside of the transom. This should extend from side to side, to spread the load, and if knees can be added to take the stresses around to the side of the hull, so much the better.

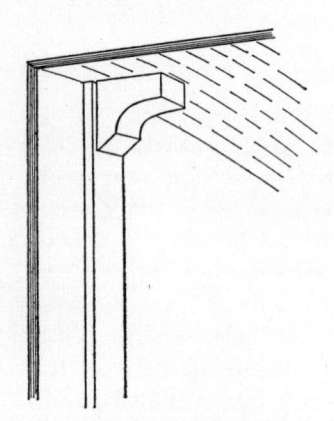

Transom stiffened with beam and knee. Both should finally be encapsulated for strength and against rot.

A rudder can be home-moulded but, unless of the thin-bladed type, must be hollow to keep the weight down. One way of making one is to laminate over a mould of thin plywood. Or you can mould the two halves separately and bond them along the join with resin and tape.

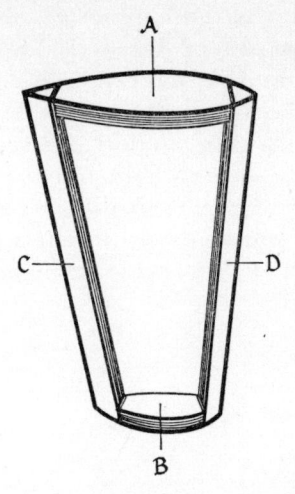

Mould frame for rudder. It can be left as part of the moulding or each half can be moulded separately, and both stuck together with glass tape and resin. A & B are profiled sections cut from plywood. C & D are battens shaped to match. All is glued together and thin ply sheeting stuck over all.

Neither method is wholly satisfactory because you run up against the problem of attaching stock to rudder. Tangs are needed, either internal or external, welded to the stock and integrated with the blade by bolting and bonding. Rather than tackle such a job I would prefer to make the rudder from timber or plywood, to be sheathed with 1oz mat and then surfacing tissue. A friend of mine had serious trouble when his rudder parted at the seams and broke away; this was a commercial model from a reputable firm.

SAFETY FITTINGS

Fittings to which you can clip harness should be placed at strategic points around the deck. Large eye-bolts are suitable to bolt on deck and coachroof, and I always fit a ringbolt into the front wall of the cockpit into which I can clip before emerging on deck. All such fittings must be really stout and securely bolted through large backing pads, with plenty of mastic against leaks as they will take many strains. If you equip your craft with lifelines for hard conditions, their anchorage points should be similarly treated.

10

Below Decks

WIRING AND PIPING

It is advisable to fit all wiring and piping before making a start on the interior appointments. If you leave it till later, you may have to bore through bulkheads, get behind settee backs, through locker sides and generally undo work you had finished off. Both wires and pipes are unsightly and should be concealed as much as possible. Pipes can run beneath the cabin sole, low under berths and so on, but I like to have wiring high up out of possible water loose below. This means leading it along the inside of lockers and shelves fitted up against the hull.

Really stout and impervious wire of heavy capacity will cost a lot but be worth it in the long run, for it will last the life of the boat—accidental damage excepted. It can be run through conduits, which are ideal but both clumsy and expensive; moulded into position; tacked in place with strips of wet tape or mat; held in clips bonded to the structure. I see nothing wrong with using simple tacking where the wire is out of sight, provided the tacks are close enough together to stop it sagging.

Where it emerges into the public gaze, there is much to commend covering it with a moulding which blends in with the decorative scheme. Resinglass is a good insulator, as long as it is dry, so that joints (which must be *soldered* if no junction box is used) can be immured in the material.

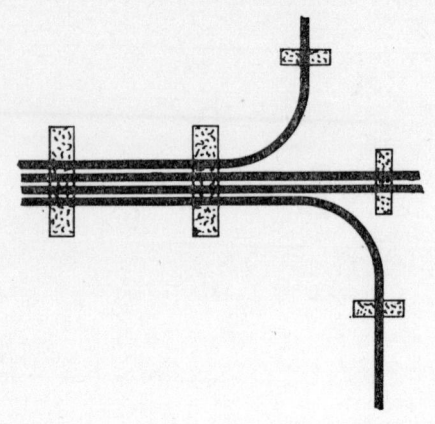

Wiring harness stuck in place with resinglass tacks.

Battery leads and wires leading from the engine and generator are best held in place with clips for ease of replacement and maintenance.

Any wiring taken through decks should be led through tight grommets or, better still, to end in a weatherproof plug and socket.

Domestic piping for water is usually of plastic nowadays, which is cheap, strong and hygienic but will not adhere to resin. If you want to keep a length from sliding in its fittings, take a very firm seizing of nylon twine around the portion to be held. It will grip the pipe immovably and if you soak it with resin mix will hold a tack securely. Such piping can, with advantage, be moulded in completely against damage, although cleaning has to be considered if you do this, but it is inadvisable to embed pipes carrying flammables like gas or fuel. These should have regular inspection and maintenance.

TEMPLATES

It is not sensible to cut out wooden components on the basis of scale drawings and the use of templates is recommended. These can be made of cheap hardboard, to be used around the house for innumerable purposes afterwards.

If you possess drawings it is an easy thing to scale them up and cut templates to the measurements. If not you will have to use the empiric method of cut and come again. This entails taking the approximate shape of a template off the hull itself and transferring it to the hardboard.

A useful dodge for this is to get a few lengths of thick solder stick from an ironmonger and join their ends with a hot iron. This provides a fair length of soft metal which can repeatedly be bent in and out of shape. Or you can use the scarcening lead-covered house-wiring which lies around in whirls and coils at most scrap-yards.

Lay this implement against the station marking on the side of the hull and bend it to follow the curve, marking the top and bottom of the section being dealt with. Carefully lay it on your sheet of hardboard and run a pencil along its convex edge. This will draw out the rough shape of the section of bulkhead to fit between the points marked on the hull. By repeating the process you will arrive at the full length of the edge of the template. Cut along the line with a jigsaw. Similar action is called for where the top of the bulkhead is to fit against the deckhead or cabin roof.

However prepared, the template now has to be offered up to its station marking. If it fits to within a quarter of an inch all round, fine. If not it will have to be spiled; this operation is really simple and fluency comes with practice. The tool needed is a pair of drawing compasses

FITTING OUT A MOULDED HULL

with legs that are stiff to move, or can be locked in place, fitted with a soft, black lead.

The template will have to be held firmly against the hull side and at right-angles to the C L for spiling; perhaps a helper will do this. If not, a batten should be wedged between top and bottom of the hull and a vertical line drawn down it—use a spirit level to find it. The template can be clamped on the batten, hard up against the hull on its inboard side. Run the spile down the edge of the template, keeping the compass point against the hull and the pencil point *absolutely level* with it throughout the operation; the reason for this emphasis is that if the spile is other than horizontal, the adjustments to the template will not fit properly—the sketch shows better than words.

As long as the spiled line passes close to the point on the edge of the template which is farthest from the hull, the amount to be taken off will be minimal. Once cut to the spiled line the template should be offered up; it should fit well but small adjustments can be made as necessary until a good fit is achieved. This need not be precise as small gaps will be behind the bonding laminations and, as said on page 29 the bulkhead should not fit the side closely.

Before moving the template, mark it top and bottom so that a vertical line can be drawn down its inboard side. If clamped to a batten, it will only be necessary to put a mark top and bottom to correspond with the vertical marking on the batten. Otherwise use a spirit level. If the line is central, the template will represent a complete half of the bulkhead. If not, the template will represent a section of bulkhead to which others will have to be butted closely for strapping.

Now offer your template up to the opposing station marks. If it fits, rejoice and mark it clearly 'P & S'. If not,

Spiling—right and wrong.

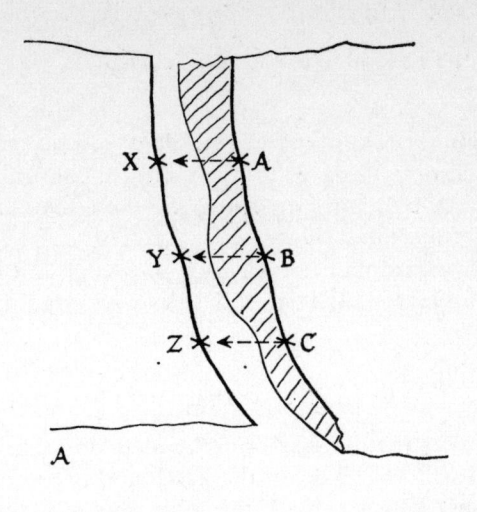

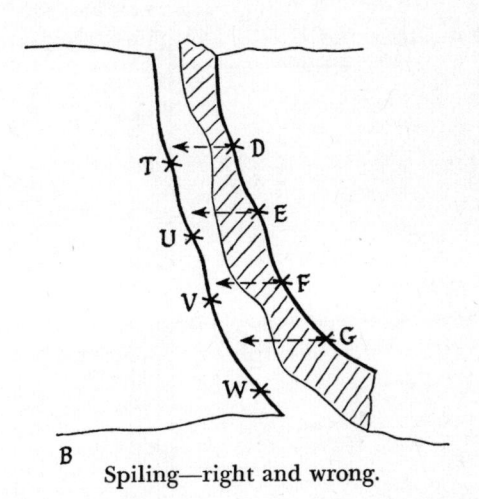

A: When trimmed, the template will move horizontally so that A, B and C meet X, Y and Z respectively.

B: Here the spile has been held at right angles to the hull. When the trimmed template is moved horizontally, D, E, F and G will not register with T, U, V and W.

and this may often be the case as mouldings are far from symmetrical, prepare and fit a second template. Mark each one 'P' or 'S' to avoid future confusion. It is a mistake to economise by using the same template with two sets of markings on it.

Templates are used for all sorts of fitting, but making one for a major bulkhead will help you sort things out for a start.

FITTING A BULKHEAD

If the superstructure has been bonded on, a full-width bulkhead will not go down the hatch and has to be made in sections. The essential thing is to make templates accurately for each section of such a size that they will enter the hatchway. By using battens and sticky tape, put them all together and try the completed template for fit.

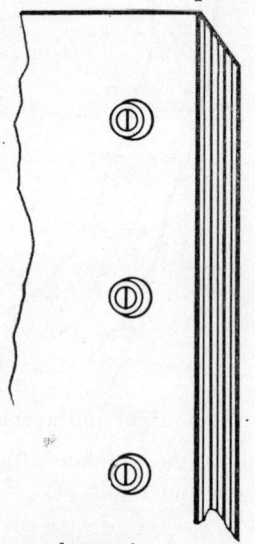

Screwcups. A neat alternative to counterboring and plugging.

Ensure that all butts are a nice fit. Then stand back and see where the joins will come; ask yourself if this is acceptable for the bulkhead itself. If not, you will have to re-section or redesign the template.

Sections of plywood bulkheads are usually joined with full-length butt-straps, especially if they are strength members, and these are glued and screwed. You can often arrange for the screwheads to be covered with framing or trim around door or access aperture. If not, I like to set their heads into those smart brass cup things. However, I run ahead.

The satisfactory template must now be laid on the ply to be cut. There are economical and extravagant ways of setting out such patterns and (with 18mm ply costing about £15 a sheet in 1973 and rising all the time) it pays to jiggle around until you have allowed for as little waste as possible. Draw scrupulously around the template and cut the wood carefully with a jigsaw.

Up forward bulkheads of thick ply may need to be chamfered:

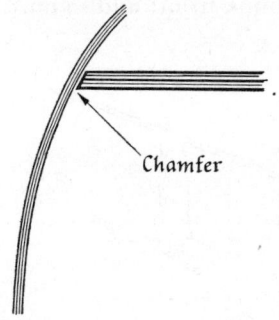

Forward bulkheads need to be chamfered to take account of the pitch of the topsides.

but this is not absolutely necessary as soft wood or neoprene can come between hull and bulkhead. If you intend to leave an air-gap it might be too big on the after side if the edge of the ply is left square.

The side sections of the bulkhead have then to be put into position and held immovable until bonding is completed. More than one pair of hands is needed at this juncture. A patient improver can be persuaded to hold on until a couple of really rapid-setting tacks are put on to bulkhead and hull. You can shore the section in position, use stout sticky tape, or, as I usually do, drive a couple of thin wedges into convenient spots.

Bonding should be done as described on pp. 27-32 and it is then relatively easy to butt-strap the remaining sections into place.

STRINGERS

I use this loose term to denote any longitudinal length of (usually) wood bonded to the hull. A stringer will, for example, carry the outboard edge of a bunk top; form or support the bottom of a shelf; act as the framing along the bottom of a bunk front; and so on.

Stringers can be secured by tacks or full-length bonds.

Any stringer fitted down on the bilge should not be sealed off with bonding, but a couple of gaps should be left along its length to let water drain down to the sump. Where the inboard side of a stringer is exposed to view, and may in fact form part of general trim or decoration, it should be bonded on the hidden side only. Bonding can consist of tacks or a full-length strip according to the demands of the job.

FRAMING

The drawing shows the front part of a bunk. The bottom frame is a bonded stringer; the others have been glued to the ply front to stiffen it up and allow less heavy, and so costly, material to be used. Do not overlook this economical dodge; quite thin ply will serve for many moderate strength purposes as long as it is surrounded with a frame. The sections of such a frame should be

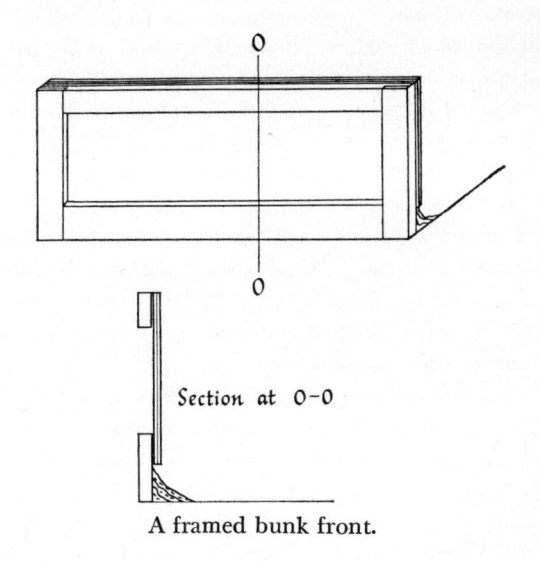

A framed bunk front.

half-jointed or otherwise integrated so that the frame is really independent of the ply, which is only there to cover up the hole.

The bunk front can be fixed to a bulkhead in many ways, the easiest way being just to bond it on.

You may prefer to use squared timber, glued and screwed to both front and bulkhead, in which case the front support for the top of the bunk can be half-jointed into it at the top. Or just glued and screwed. In this case do the joining before attaching the framing so that screwheads will be covered. I use whichever method of joining up is more sensible for a particular task; bonding is easy but may not be neat enough for some exposed joins.

GALLEY

A galley top and its surrounds are best covered with Formica or other hard, decorative plastic. This gives protection against hot splashes and is easy to keep clean. It can be stuck on wood or resinglass with Evostick. Smoke stains are a problem around a galley area, particularly overhead, and you must prevent heat from the cooker, perhaps a flare-up, from affecting the hull or even setting it alight. You can insulate the roof from damage and stains by fitting a sheet of asbestos above the galley, leaving about an inch of air space between the two. Its exposed side can be covered with a sheet of aluminium, copper or stainless steel for appearance and ease of cleaning. Fix the sheets with suitable screws entered into small blocks of wood bonded on.

A galley sink, unless pumped out below water, should drain outboard a little way above the waterline. Fit a seacock to stop inflow when adversely heeled.

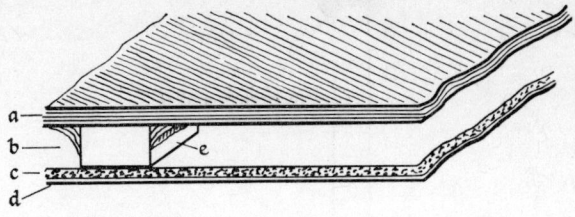

a Deckhead
b Airspace of about one inch
c Rigid asbestos sheet — half inch
d Thin copper or stainless steel sheet
e Small wooden blocks bonded at intervals.
   Metal and asbestos held to these with screws

Fireproofing the galley.

LOCKERAGE

You cannot have too many lockers aboard and numerous small ones into which gear and stores can be firmly chocked are better than a few large ones. You may have to design for a hanging locker and an oily locker if you want to spend much time offshore. It is relatively easy to mould small lockers, complete, and bond them in but wooden ones—or at least wooden doors—are attractive. A locker door should hinge down and not up, so that it can always be seen to be either open or shut.

Cockpit lockers sometimes have wooden or resinglass lids supplied loose or fitted, but you may have to make them. Hinging on mouldings should be strengthened, as explained, and as the lids serve as seats the job must be a secure one. Lids should fit down on weatherproof gaskets.

Apertures may have been cut in cockpit walls to give access to stowage space, or you may decide to cut them out yourself. They will need waterproof and easily securable lids. It pays to mould or bond a flange around their top edges to deflect water.

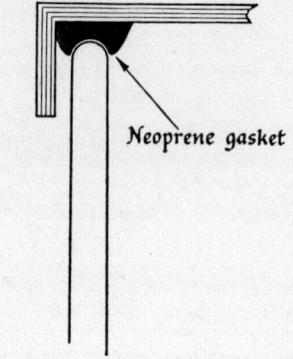

Weatherproofing cockpit lockers.

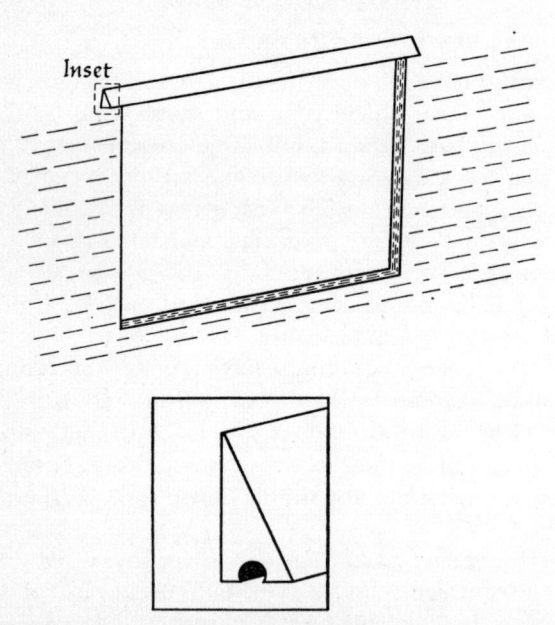

Inset

Keeping water out of vertical apertures. Note groove in bottom edge of flange which stops water running under.

BELOW DECKS

A wooden or moulded lip should be fitted inside the aperture on which the lid can sit. This can be prefabricated and held in position with sticky tape, or clamped, while the glue or bonding sets.

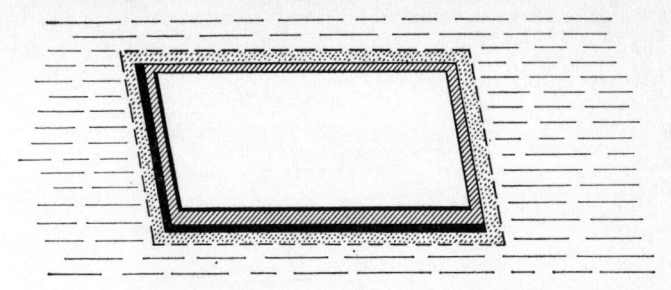

Lip fitted inside aperture to accept lid.

A simple turnbutton at either side will hold the lid securely in place if it is biassed to squeeze down when it closes.

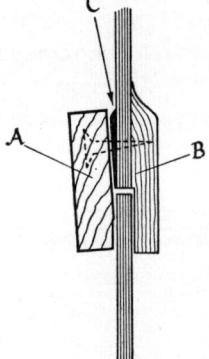

A biassed turnbutton.
A: Wooden turnbutton.
B: Lip (see previous diagram).
C: Thin wedge fixed to hull. The turnbutton pivots at right angles to this and so exerts squeeze on lid when closed.

FITTING OUT A MOULDED HULL

CABIN SOLE

This can be a module but is better made of ply, tongue-and-grooved strips or closely fitting fore and aft planks of about three inches wide. Teak is admirable to lay in planks. If there is a sump beneath, you should rest the sole on wooden floors.

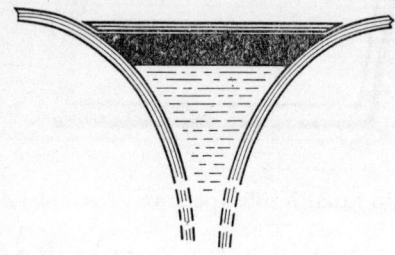

Cabin sole resting on stout wooden floors bonded into keel cavity, or bilge.

If you intend to stow clumsy gear like anchors and chain in the sump, floors should be removable and simply rest in beds moulded to receive them. If not, they can be bonded in. In either case skin them with tissue and mix or give them a good coat of paint; the bilge is mostly wet.

With bilgekeelers especially, it is possible to make the sole of stout (about $\frac{3}{4}$-inch) ply and butt it close to bunk fronts or other stringers on each side of the walkway. Preferably removable for cleaning out the bilge, it can be screwed with small angles to the stringers or to pads bonded underneath it. Its own weight would keep it in place without any fixings, of course, but it could come adrift in bad conditions. A capsize is extremely unlikely for most of us, but if there are heavy objects beneath a loose sole they should be kept captive. Remember this consideration if fitting loose floors under a sole.

BELOW DECKS

THE HEADS

A pumped toilet needs seacocks for inlet and outlet and they should be easy of access. The pan should be mounted on a stout wooden block bonded on. The bolts can be glued into the wood with Araldite and the base of the machine secured with washers and nuts. Use stainless ones if possible against corrosion. Pan and block should live in a little box with a lid, and this should be lined with smooth gel-coat or lead. Unless this precaution is taken it will be hard to clean and any exposed wood will get contaminated and smelly, even if painted or varnished.

Chemical type toilets need similarly to be firmly anchored if you are going to use them at sea. Although there will be no frantic pumping, the weight of a body being heaved around in a seaway can exert quite a force.

CABLE STOWAGE

Chain usually leads through a pipe on the foredeck and down to a locker in the bows. This should be lined with wood on frames to cut down on noise and damage.

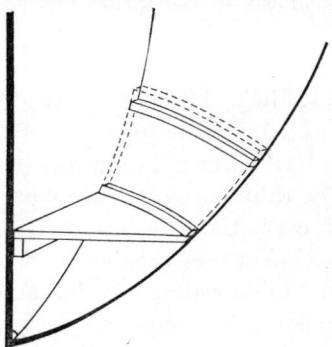

Section through chain locker. Plywood base: dotted lines show ply panel secured to bonded battens: note limber hole at bottom.

A chainpipe needs stout backing pads as, if the chain jams on run-out, stresses could be large. It is sensible to fit as wide a diameter of pipe as is acceptable to you against this eventuality, and there is not much clearance if you take the ones recommended for a given size of link. It is difficult to stop water getting down a chainpipe, and the following simple expedient is very effective:

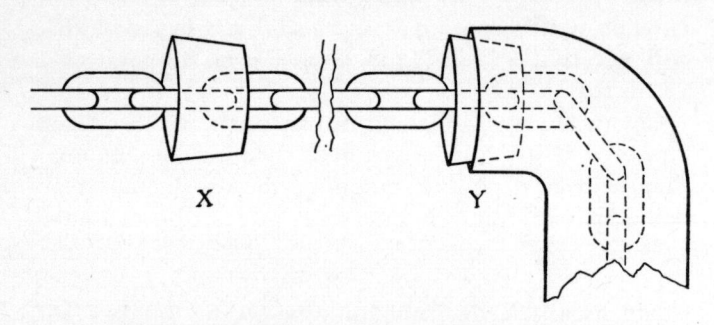

A plug of flexible resin is moulded around a couple of links, (X); when chain runs below, plug fits into pipe as shown by dotted line, (Y).

It permits the chain to be instantly ready, which is not so if it is unshackled and the pipe corked up.

Anchor rope has to be kept on a reel to be of practical use. This can be dismountable and taken to and from a holder on deck every time you use it. It's suicidal under bad conditions, but a reel can be fitted below between two longitudinal bulkheads in the forepeak and the line led down through a rope-pipe on deck. Even in good conditions, this is not a very practical proposition for a shorthanded craft as the line has to be taken in from below.

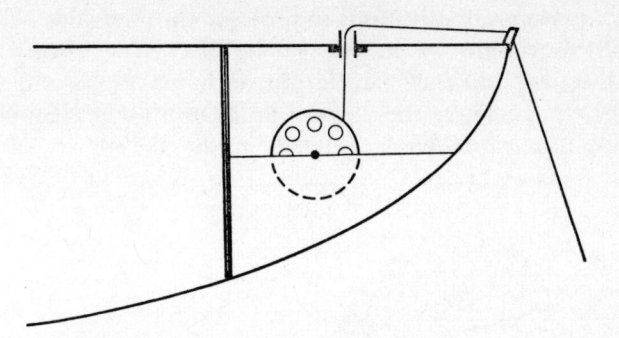

Cable reel mounted between fore and aft bulkheads in forepeak.

A reel holding a sensible forty fathoms of stout nylon line is a very bulky object to fit, with holder, on a foredeck or at the foot of the mast. Its anchorage will need to be massive because the leverage exerted by a fouled anchor or heavy seas will be great and could cause serious and dangerous structural damage to the boat.

VENTILATION

This matter is sadly overlooked in the manufacture of many hulls and you will usually have to set about designing your own installation. Dorade boxes and flat vents have been discussed, but remember that it is essential to keep a boat aired out thoroughly at all times. It is also necessary to vent fuel tanks and engine compartments against smell and risk of explosion, and galleys and heads for hygienic reasons.

Fuel tanks and engine compartments, also lockers holding fuel or gas containers, should be fitted with inlet and outlet vents. The latter should lead to outside the hull, through a pipe if necessary. Pairs of vents should face in opposite directions to promote through-flow.

It is simple to mould small shell type cowls as illustrated, and they can be put on hatch covers and in other spots. There should be a lip to stop water creeping in and a gauze mesh bonded in the mouth to keep out spray and rain.

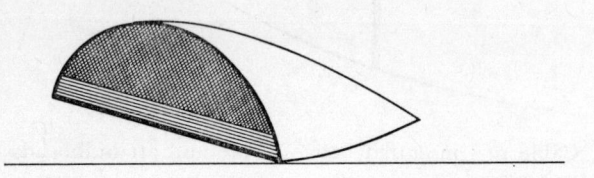

Easily moulded shell-type ventilator cowl.

TANKAGE

I am not a proponent of built-in tanks and prefer to use loose containers for water and fuel. These can be easily ferried for refilling, replaced if damaged, cleaned regularly and, if kept in a locker fitted to hold them, will not move about or come adrift. However, you may wish to have permanent tanks, and a method of moulding one in place is given on page 85.

Hollow fins are natural repositories for moulded-in tanks, which keeps weight low and makes access to their top covers easy. There is a possible risk of contamination by bilgewater. Tanks can also be sited under bunks, at the base of lockers and so on, and then have, in many cases, filler pipes and vents leading to outside.

Inlet and outlet pipes are preferably fitted with small skin fittings into the tank and not moulded into its material; leaks can occur at such places and integral pipes are hard to replace. Every tank should have a removable cover large enough to let you get inside with your hands, rags and pump when it needs cleansing. This should be compressed on a leakproof gasket of something like neoprene, but check that it would not be rotted away by fuel. The best way of holding the cover on is with bolts which thread into captive, or embedded, nuts.

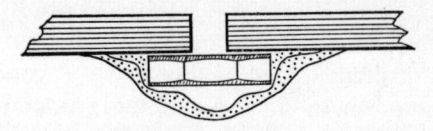

Embedded nut. When embedding, the bolt should be well greased and screwed into place. The nut will be immobilised, but the bolt will unscrew easily.

Tanks holding more than three or four gallons must be fitted with baffles to stop their contents swilling around. You can make these of plywood, but resinglass flats are unaffected by water or most fuels. Each baffle should have a drainhole at the bottom to allow the liquid to remain level and reach the outlet tap or pipe. At the top they should reach close to the bottom of the cover.

DECORATION

The more of the unattractive interior resinglass that can be disguised, the less like a hairy bathtub will you find your boat. It would take a book in itself to deal with the vast subject of decoration, touched on previously in these pages.

CONDENSATION

There is one point of real importance to be discussed—the problems of condensation.

A resinglass boat is an ideal condenser—much more than a wooden one whose bare timbers absorb quite a bit of damp. Topsides may be exposed to a warm atmosphere, even during an English summer, and will almost always be at a higher temperature than the immersed hull section. On these lower surfaces the air deposits some of its water content and the sides are nearly always wet to the touch. Bilges stay moist or awash. In cold, damp weather an unheated boat simply drips all over.

You can take advantage of the system of condensation to keep damp where it is out of the way and more or less under control.

Ideally there should be an airspace between the hull and its interior decoration, down which drips and moisture can run. Thin ply stuck on battens is a practical way of achieving this end.

Wrong   Right

Thin lining plywood fitted over battens as shown. Wrong to join them horizontally as water will condense on top and have no escape. If sloped and unjoined at one end water will run away behind the lining.

Superior hulls often have a lined coachroof; underneath the moulding is placed a layer of soft stuff like polystyrene foam, which is then skinned with a thin layer of resinglass. This effectively reduces condensation to a great extent, and you can do something similar with your boat if it is not so treated.

Line the inside of the living space with polystyrene tiles or sheeting before sticking on the vinyl or other lining. Some decorative sheeting can be bought with a foam back, and this serves the same purpose and saves one application of adhesive. Or you can use a special anti-condensant paint which contains granules of cork; this is most effective. Anything can be used which will break up the (relatively) smooth surface of the resinglass, and at the same time give heat insulation, will reduce condensation.

Having made every effort to stop water forming on exposed surfaces, make sure that sufficient areas of untreated hull are left beneath bunks, inside lockers and so on to permit the inevitable amount of condensation to take place. As I explained, spaces should be left under stringers and frames for it to percolate into the sump.

WINDOWS AND OTHER LIGHTS

Apertures for windows and portholes have to be sawn out of the hull. My boat has cavernous quarterberths, always full of gear which is tedious to rummage in gloom. I have fitted small windows to the sidewalls of the cockpit to light them up.

I strongly advise against using plastic or rubber channeling to set windows in, as is so often seen since boats started to be sold as floating caravans. They can be stove in by a sea, or just drop out when the boat flexes under stress.

Windows can be made of acrylic or polycarbonate sheeting cut to size, or pieces of laminated glass. The armoured, shatterable glass used for windscreens would, if available, be wholly wrong to use. The normal way of fitting windows is to bore them to accept a close line of countersunk bolts around their perimeter and bed them on mastic. This must always be pliable and the use of resin is unwise. The boltheads can be covered with a wooden or moulded frame, although it is only necessary for reasons of appearance; many folk do not bother. Shanks and nuts inside the cabin can be disguised with frames matched to your general scheme.

It is of advantage to fit opening portholes, of a sensible size and pattern, in the region of heads and galley. They will have inside and outside frames to pull down with nuts and bolts on to mastic bedding. Opening windows of the sliding type are prone to leaks.

I like to keep windows very small in the interests of safety and to avoid cutting great weakening holes in the hull. It is easy to get plenty of light into the interior by using translucent resinglass for the top of hatch covers. This is better than inserting pieces of Perspex or glass, because the cover will be integral for strength and impervious to leaks. If moulding your own coachroof, you could include areas of translucency in the roof if wished. Windows can be fitted in washboards or doors, or you could mould translucent washboards and frame them with wood to take the rub of the slides. My own top washboard (of two) is fitted with a small louvre, about ten by eight inches, whose slats are of translucent $\frac{3}{8}$-inch Perspex. Light and air can both enter and I use this top section at moorings. A solid replacement lives below for use, if needed, in battening down afloat.

Incidentally, Perspex is troublesome to saw with hack-saws and files, but one of the hard cutting discs used in an electric drill is very effective. You can follow a scribed line most accurately because the disc will flex enough to go round curves. The same disc, when used

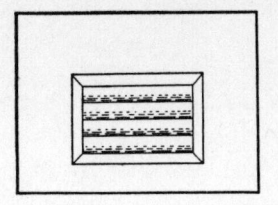

Louvres in washboard provide both light and ventilation.

on a rubber backing pad, is ideal to otherwise trim and shape the resistant acrylic. If acrylic material is well roughened, you can bond it securely with Araldite or epoxy resin to wood or resinglass.

NON-SLIP DECKS

Many decks are moulded with a pattern of raised dots, or diamonds, in an attempt to make them skidproof. This is not outstandingly effective because the surface of the raised areas is still of gel-coat and slippery when wet. You can improve matters by painting the areas with a non-slip paint or patent anti-skid preparation. Outline the areas with masking tape before applying the paint and the effect will be neat, especially if the paint is coloured to tone with or contrast with the untreated surfaces surrounding them.

Alternatively, you will find it cheaper, and perhaps better, to mix a fair amount of fine cork granules into ordinary polyurethane paint. Do not use sand for the purpose. It is abrasive to bare feet and will scuff through the soles of deck shoes and the knees of oily trousers. Sawdust is too fine and soft to use, and tends in any case to clump up and be difficult to mix in evenly.

11

Engines

What is said below applies to both engines for powerboats and to auxiliaries in sailing craft. The more powerful the engine, the stronger will you have to make its installation, but the principles are the same.

There is little to be said about outboard fitting which is not obvious. Slides, brackets and other fittings on the stern into which the motor will be placed should be bolted through strong backing timbers, in similar fashion to rudder hangings. Motors should never be clamped directly on resinglass, but battens moulded on to take the compression. If the lower part of the shaft may come in contact with the hull, a small hardwood pad will protect the threatened area. Make quite sure that any hot exhaust pipe cannot be swivelled to touch the hull.

INBOARDS

First consider the physical problems of getting a motor down through the hatchway, having good access to it for running repair and regular maintenance, and possible future removal for replacement or major overhaul. Very often a hull is so designed that if the engine is fitted in the small space beneath bridge deck or cockpit sole, it is pretty inaccessible for practical purposes. It can often be better to site it six inches or even a foot forward of the designed position so that it is clear of obstruction. You will have to make a corresponding adjustment to the length of the propeller shaft, which needs to be thought about if you buy engine and stern gear as a package.

You will need access to the stern gland for greasing, and *don't forget a driptray*! A sensible tray should extend well forward, aft and on either side of the bottom of the engine so that no oil, fuel or other muck can get into the bilges even when the boat is heeled. It is best detachable so that it can be taken out and emptied, but at least there must be provision for draining or pumping it out. Remember that engines have to have their oil changed, and the drainplug is better sited over a driptray than an open bilge. Obviously, the tray must be big enough to accept a sump-full of oil.

Controls need careful consideration for attaching and siting. They should be accessible to the helmsman, who may often be alone on deck. Remote controls are best of the mechanically-linked type, as Bowden cables can corrode and jam if greasing is not regular and conscientious.

BATTERIES

Batteries of sensible capacity should be carried, and such weighty objects should be stowed as low as practicable. I dislike putting them in the fin; water gets everywhere and a mixture of seawater and battery acid gives off chlorine, which is highly toxic. Wherever you site them —and under a quarterberth is as good a spot as any— remember that a starter takes a heavy current and battery leads should be both short and of as adequate a capacity as is called for. Resinglass is unaffected by acids, so a box of this material could be moulded in to take a battery. If you choose NIFE or NICAD batteries, they are fed with an alkali which must be kept away from resinglass. Some form of strapping for anchoring batteries should be fitted, for they cause havoc if adrift in a seaway.

FUEL TANKS

I have already expressed my dislike for large, built-in tankage. If fitted, it should be isolated by a watertight bulkhead so that fuel cannot possibly leak into the bilge. My boat has a fueltight locker aft, partitioned to accept fuel cans. On the front wall of the cockpit I have moulded a four-gallon tank; the pipe is fitted with a tap underneath the tank, out in the cockpit. When this is shut off and the carburettor run dry, there is small likelihood of a fuel fire below. The small inconvenience of transferring fuel is well worth the trouble.

ENGINE BEARERS

The function of engine bearers is to carry the dead weight of the machine and resist the stresses created when it is running, principally torque and vibration. They should be virtually an integral part of the hull, so that stresses are spread widely, and stout enough to remain rigid at all times. Bearers which flex, or an engine which is not firmly mounted, will cause misalignment of shafting and consequent wear and loss of efficiency. It is possible to mount the engine on flexible mounts, in which case vibration will diminish, but this has to be compensated for by fitting a flexible coupling between engine and propeller shaft; possibly an additional thrust bearing. It is advisable to read about problems of engine

ENGINES

fitting and lining up, and this chapter is only concerned with the resinglass aspects.

Bearers can be of many types, a couple of which are shown.

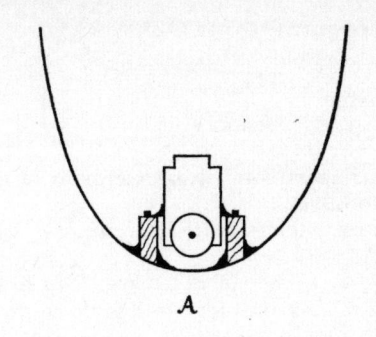

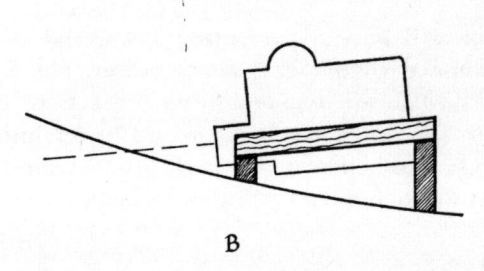

A

B

A: Fore and aft trunnions bonded directly to hull.
B: Trunnions set thwartwise with bearers running fore and aft.

Thick laminations should be used to bond bearers to hull, preferably as thick as the hull itself if bonded on one side only, and half that if on both.

Resinglass hulls flex to a surprising degree, and you need to extend the bearers as far as possible aft along the length of the shafting.

FITTING OUT A MOULDED HULL

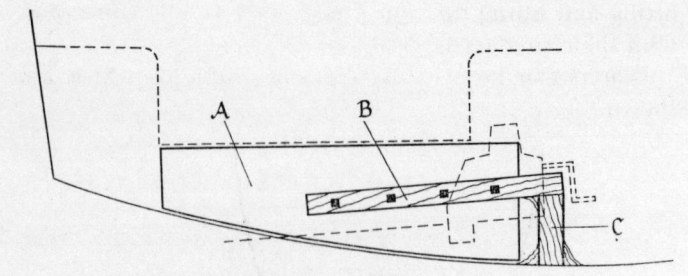

A: Plywood stringers bonded as extensions of cockpit sides and to bilge.
B: Stout timber engine bearing trunnions, bolted and glued to A.
C: Upright, stout timber blocks. These help to keep B in place and also transfer stresses down to the bilge. Bonding should be spread over a wide area; or a transverse timber can be interposed between uprights and bilge.

An engine will have to be bolted down, and so if you have laminated completely over a bearer, put a gasket of neoprene beneath the mounting brackets to prevent damage to the resinglass. Bolts should be set into wood or through metal, if you use angle iron as shown, and never just through a lamination of resinglass.

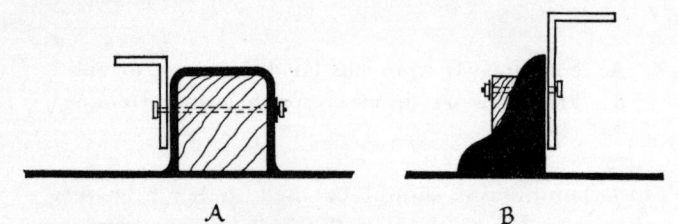

A: Angle iron bolted to channel formed around timber.
B: Angle iron bolted to resinglass flange. Note wooden backing pad to spread load and prevent crushing.

ENGINES

STERN TUBE

You may have bought a hull with the stern tube moulded in, but if not it is a simple matter to make one. Its inside surface should be smooth, so that your tube can best be moulded over a metal tube or rod covered with polythene sheeting. If you grease the metal before rolling the sheeting around—two turns at least—the

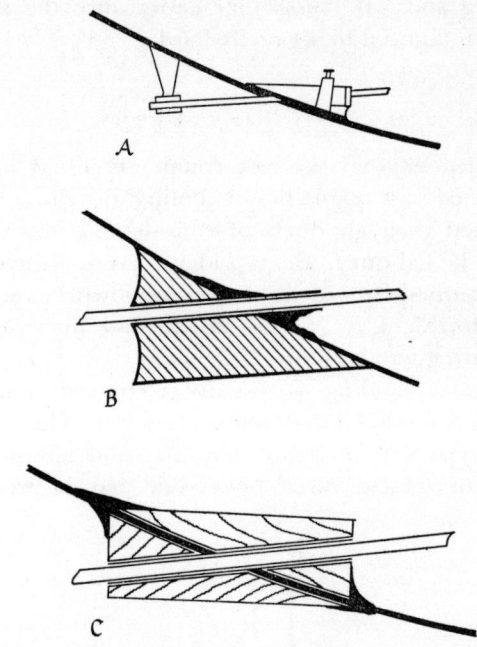

A shows adjustable shaft log which allows of great latitude in lining up.
B shows resinglass stern tube heavily bonded to inside of hull and led through skeg outside.
C shows shaft log made from timber block drilled and split. Stern tube taken through this. Timber heavily bonded into place on both sides of hull.

FITTING OUT A MOULDED HULL

former can be knocked out after the moulding has set.

There are several ways of taking the propeller shaft through the hull.

If you mould in your fabricated tube, take care that it is accurately lined up and held so during the process of bonding it in place. Pass a length of closely fitting dowelling through the tube; this can be clamped to battening inside and to a possible A-bracket outside. Grease it beforehand so that it cannot adhere by mistake to the bonding and will knock out easily once the moulding has been allowed to set really hard.

COOLING SYSTEMS

Air-cooled engines are not commonly fitted nowadays. They need vast quantities of cooling air which has to be exhausted through ducts of considerable size. As these have to be led out to the topsides, there is always a likelihood of quantities of water getting down below, despite efforts to shield the outlets. Inadequate ducts will cause overheating and trouble.

Sea-water cooling is usually employed, and enters through a seacock fitted below the water. This should be of the type with a removable filter and mounted, with plenty of mastic, on a bonded-in pad of wood. The

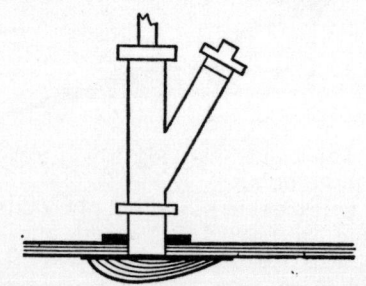

Grid fitted over inlet for cooling water.

ENGINES

throughbolts can sometimes be cajoled into holding the grid needed to cover the inlet hole outside the hull. Don't omit this grid, or your system will inevitably get choked with small weed and mundungus.

Cooling water should always be exhausted above sea level, and if not used to cool the exhaust as well can be led through a skin fitting quite high up where you can glance at it periodically to ensure that it is circulating properly. A secondary circuit may be provided to take the hot cooling water through a copper pipe which runs round an oily locker, or even round the cabin sole. It is satisfying to get back some of the heat you have paid for.

A cooled exhaust usually goes out a little distance above the waterline because it has to slope downwards away from the engine. In this case, you can fit a three-way valve in order to check on circulation.

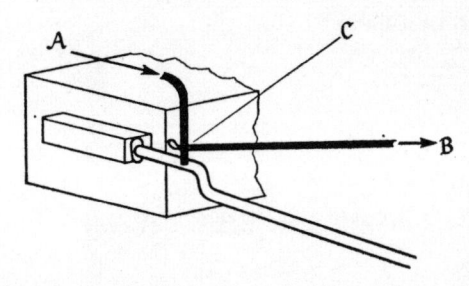

Cooling water fed in at A can be directed to exhaust cooling or, by means of three-way cock C, be led out at B and over the side where it provides visible confirmation of the circulation of cooling water.

EXHAUST PIPES

An uncooled exhaust pipe gets very hot and you must take extreme care to insulate it against risk of fire and damage to the resinglass of the hull. Where it runs from engine to outlet, it should be wrapped with thick asbestos, either rope of about $\frac{1}{2}$-inch diameter or $\frac{1}{2}$-inch sheeting held on with Jubilee clips at sensible intervals. Space it well away from hull, bunk tops and so on so that in addition to the asbestos there is a good air-gap all round. It should never pass through a compartment containing flammable substances.

A water-cooled exhaust is only very hot up to the point where water is injected, and I have seen examples of economical thinking resulting in the pipe beyond that point being scantily insulated. This is shortsighted; a water stoppage could cause the pipe to heat up rapidly and start a fire. It is more sensible to wrap it as securely as an uncooled one, for its full length.

Where taken through the hull, an exhaust pipe needs to be really well lagged with asbestos. This can be skinned with resinglass on to which a bonding can be made in the form of a flange.

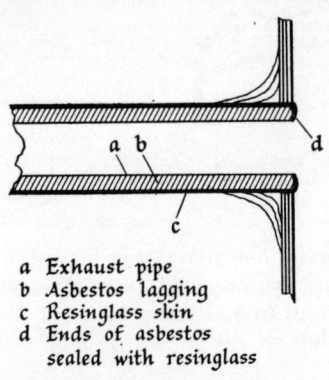

a Exhaust pipe
b Asbestos lagging
c Resinglass skin
d Ends of asbestos sealed with resinglass

Insulation of exhaust exit.

Seal the outboard end of the lagging against soaking up water.

ENGINES

FUEL PIPES

These must be kept away from the exhaust for obvious reasons. Annealed copper piping is usually recommended, with a coil made in it at some point to allow of expansion, but I cannot see that there is anything against using reinforced plastic piping as long as the correct fittings are used at each end and it is kept away from heat which would soften it and cause petrol to vaporise, expand and possibly burst the pipe with disastrous consequences. It can conveniently be tacked in place.

FIRE PRECAUTIONS

Ideally, an engine should be in a compartment which is internally insulated against fire with glass wool or similar material. There must be a vent for fresh air to enter and an outlet for hot air and fumes. Powercraft are often fitted with an exhaust fan, which must be of the spark-free type, so wired that it is not possible to start the engine until the fan has run for a short time. This ensures that petrol fumes cannot be ignited by starting currents. Carburettors should be fitted with a flame trap over the inlet.

Even with all these precautions, it is imperative to carry extinguishers which can be got at instantly in the event of trouble. Foam or dry powder types are preferable to others, although BCF is the most popular extinguishant today. Avoid the obsolete trichlorethylene models which are to be picked up cheap at auctions and elsewhere. This chemical can emit toxic fumes.

Starting and stopping routines will go a long way to prevent fires. If the main fuel cock is above decks, it can be turned off and the pipe run dry thereby ensuring that there is no risk of a fuel leak below decks giving trouble.

When starting, it is advisable to open the lid of the engine compartment to make sure that the carburettor is not sticking and causing petrol to overflow into the bilge. Any spilled petrol should be mopped up and fumes allowed to dissipate before the engine is started.

Diesel fuel is not highly flammable but should be cleaned up if spilled; it smells so horrid that this will probably be automatic.

Fitted tanks should have their filler pipes leading down from the deck. Never fill the pipe itself, as the pressure transferred to the tank will be high and might cause a leak or burst.

12

Miscellaneous

COCKPIT DRAINS

Skin fittings should be screwed or bolted on mastic and not bonded on. There is no stress on them and mounting pads are not needed. Seacocks should be used if the outlets are below water, and in this case it is best to use a bonded block for mounting against opening and closing stresses. Drains should be crossed and it is not advisable to use resinglass pipes for the purpose, as there is always flexure of the cockpit structure under crew weight and movement. Use plastic piping attached with stainless Jubilee clips.

BILGE PUMPS

Bilge pumps should be accessible for immediate unblocking, removable if necessary for major repairs, and solidly bolted through backing pads. One should be in the cockpit where the helmsman can use it without leaving the helm, and another sited below for use when battened down. Some folk lead bilge water out through skin fittings, but it is often simpler to have a length of outlet piping which can be put over the lee deckside. This ensures that water cannot back up and run into the bilge in bad conditions.

FITTING OUT A MOULDED HULL

ELECTRONIC GEAR

Modern echo sounders and electronic logs have their sensing devices fitted on the inside of the hull, and radiate through the resinglass. If, by mischance, they are sited over an area containing air bubbles in the moulding, readings will be inaccurate. You would be sensible to site them temporarily until the validity of readings has been established, and only then bond or fit them permanently in place.

NAVIGATION LIGHTS

These can effectively be sited on the side of the coach-roof, on a Dorade box or in other places where they will be clearly visible and can be correctly screened. Provided that you can open them for bulb changes and maintenance, it will be advantageous to bond them immovably into position. Wires leading out through the hull skin can go through rubber grommets.

An overtaking light can be recessed into the transom and so kept away from damage.

GRATINGS

Cockpit sole, bridge deck and very often bare resinglass cockpit seats can be transformed with gratings. These can be made removable, but if drainage holes are left at intervals on the under side, it can be easier to fit them permanently into place. Use epoxy resin as an adhesive; glass cloth is not needed unless the curve of an area demands that filling is needed under portions of the gratings to level it off.

Gratings can be of timber, which looks good and scrubs clean. It is possible to mould resinglass slats on top of

MISCELLANEOUS

seats by using a U-shaped mould of the required lengths. The top of the slats can be finished off with tissue which gives a reasonably smooth-looking but non-slip surface.

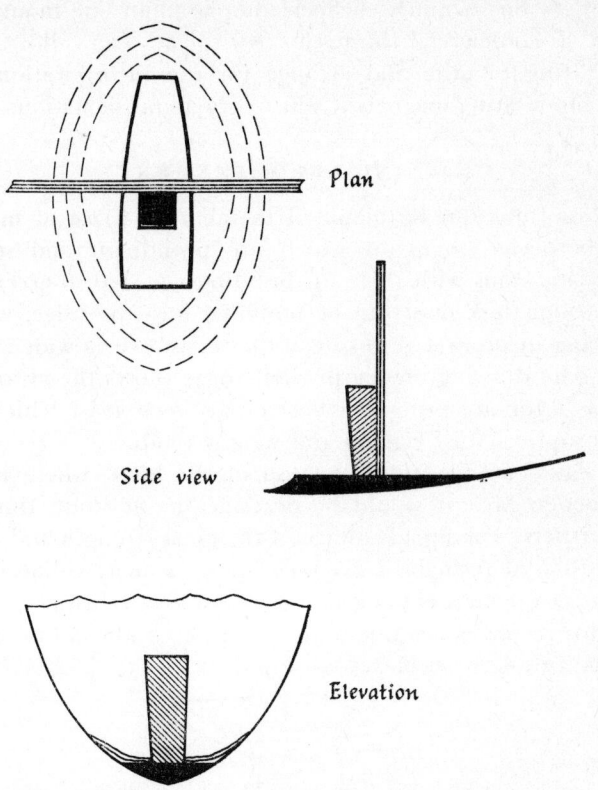

Plan

Side view

Elevation

Mast stepping arrangement. Heavy wooden step is glassed into keel 'V' after being laid on a laminate. Glassing is spread widely as shown by dotted lines, and serves to stiffen hull and spread loads without using other timbers. Tabernacle type receptacle is bonded to bulkhead to accept foot of mast. Mast lies against bulkhead, which is tied at the top to the mast collar—see Fig. on next page.

FITTING OUT A MOULDED HULL

BINNACLE

A functional binnacle is fairly simple to mould in resinglass to match or contrast with the topsides. This allows you to buy a much cheaper compass than one mounted in a commercial binnacle. Also, you can allow for lighting to taste and arrange for the incorporation of compensating magnets if you worry about deviation.

SAILING ACCOUTREMENTS

Mast steps can be moulded to take the place of metal tabernacles, many of which are prominent and seem incongruous with their surroundings. A step to accept a through-deck mast can be moulded into the bilge, webs being incorporated to spread the thrust over a wide area.

The drawing on the previous page shows the arrangement for accepting the heel of my own mast which is unsupported by rigging and weighs 160lb.

Ancillary to this is the top of the bury, where in a wooden boat it would be necessary to fit stout, timber partners. Taking advantage of the great strength and versatility of resinglass, my boat sports a mast collar into which the mast is wedged and a mast coat taken about it. This increases headroom below and a bulkhead has been tied into the complex.

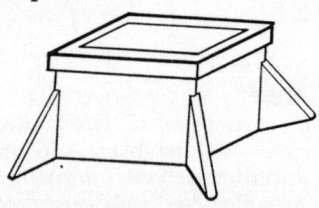

Mast collar. Moulded of heavy laminations of C S M and W C. Flanged at top so that mast collar can be secured and waterproofed with a lacing. Webs are moulded over included plywood.

13

Protection, Maintenance and Repair

PROTECTION

Resinglass is not only much thinner than wood of the same strength, but in some ways more vulnerable to damage. An anchor dropped hard on a planked deck may make a dent. This may be unsightly but can be ignored for practical purposes as long as its existence does not give rise to leaks or rot. A similar impact with a resinglass deck would be likely to crack or craze the gel-coat. While the strength of the deck would be little affected, I have already explained the dangers of letting water percolate into the interior of a moulding.

Again, toerails and cockpit coamings take much hard wear and scuffing from lines and gear brought in over them. It is easy to cut out damage and scarf in a replacement section to a wooden rail or coaming, but a moulded one will have its gel-coat damaged and perhaps worn right through over a period of time.

If you clobber a jetty with a six-inch oak stem you may lose a couple of cubic inches of timber. This does not seriously weaken the stem and can, in any case, be put right quite simply. On the other hand if you plough off a quarter of an inch from a half-inch thick moulding, weakening is severe.

Protection depends greatly on the adherence to routines which will avert impact damage—care in handling ground tackle; in coming alongside; in lifting weights aboard; and so on. Scuffing and abrading wear is best prevented by shielding vulnerable surfaces and areas with materials which are relatively inexpensive, replaceable, and stout enough to take their burdens without affecting the underlying resinglass. Cappings to toerails and cockpit coamings are good examples. A grid, slats, plywood section or sheet of thick neoprene bonded on the foredeck may be used to protect its surface from not only ground tackle, but the gritty feet of crew working there. Grit carried by the patterned soles of deckshoes acts like sandpaper on gel-coat. If a tender is carried on the coach-roof, you should interpose wooden blocks between the resinglass and the parts of the tender resting habitually in the same place. Its lashings may be of webbing or rope and, unless some form of protection is given, these will gently but inexorably grind their way through gel-coat and on into the laminations.

Silly things are often overlooked, like coils of warps being hung on hooks up forward. Synthetic ropes are tough and if they constantly swing about and rub on the hull will cause damage. Conversely, although this has nothing to do with damage prevention, if you hang clothes in a locker remember that if they swing about with the motion of the boat, they will themselves wear thin and holy.

Slats, grids and so on are best only tacked in place firmly enough to hold them, but cause minimal cutting away of bonding if they need to be replaced. It is best not to bond cappings which can be scored deeply or cut through; they should be throughbolted on backing pads.

If your boat lives on a trailer, or cradle, during the winter you should see that wooden wedges or rubber padding is put between any metal shores and supports which could come into contact with the hull. It is surprising how much a boat ashore will move about during the winter under the effect of winds. Also, people working in and about a boat cause her to move perceptibly.

MAINTENANCE

Given intelligent use of protection against wear, a resinglass hull needs very little maintenance in comparison with a wooden or steel one, but it needs some attention.

Maintenance afloat consists of keeping an eye open for minor damage and doing whatever is necessary to stop it getting worse; keeping the decks and interior clean, principally to swill away grit; and caring for wooden and metal fittings as you would if the boat were wholly of another material.

Once ashore, a resinglass hull should be cleaned out thoroughly and inspected for wear and damage. Make a note of any repairs to be done and then give the whole of the exterior a thick coat of wax polish. This need not be rubbed down and will weather away during the winter; it may need another coat toward the spring. *Avoid using silicone waxes.* It is almost impossible to remove all trace of an application of silicone and if you have occasion to repair damage or bond on, the presence of silicones will seriously affect the job.

I do not use a boat cover. Resinglass does not weather and plenty of rainwater keeps the decks clean and sweet. The hull remains well-ventilated, which reduces likelihood of rot in the wood below. However, I am lucky enough to be in a pollution-free area; where industrial dust and pollutive deposits are to be met with it would be more advisable to fit a close-fitting, well-ventilated cover over all. If you do this, guard against chafe from lashings.

In the course of time, resinglass will fade and lose its gloss. Faded pigment cannot be dealt with—blue hulls seem to be most affected—and if you are not prepared to sail in a boat which does not look pristine, it will have to be painted as previously explained. Lack of gloss can sometimes be dealt with. Hard work with finest grade wet and dry paper, used wet with bags of water, followed by a polish with good wax will often do wonders to a dulled exterior. Patent polishes abound and you can use one of these. Brasso and Perspex polish are other expedients. The main ingredient is elbowgrease. Once the gel-coat has dulled beyond saving you will have to consider painting.

REPAIR

It is very easy to repair damaged resinglass—even quite considerable damage. It is often very difficult to do this so that the repair is invisible.

A repair to the outside of a hull means that the gel-coat has to be matched accurately for colour, and that the line of the hull must be followed accurately. Colour-matching is a devil of a problem, because the usual difficulties arise as with paint or any other pigmented material; the colour looks slightly different when it has dried out than when it is being mixed and applied. I know of no easy solution. Presumably you could mix a quantity, catalyse a few drops, and try it for match; the quantity of pigment could be adjusted. Quite frankly, it is not worth while and, if you ever have to effect a major repair you will undoubtedly find it simpler to paint all over afterwards. Nothing will stand out so much as a largish area which is slightly off-colour. A small patch can be disguised with, preferably, two-part epoxy paint carefully matched for colour.

MINOR REPAIRS

Small holes and dents can be filled with putty, either polyester or epoxy. Putties can be bought ready mixed, or you can make your own from resin and additive; pigment can be added as required. Make sure that the depression is dry and winkle around inside it with coarse paper to produce a rough surface on which to put the putty.

Be careful of cracks. A surface crack can be puttied up but prod with plenty of strength to make sure that the crack is not extending right through the laminations. If so, you will have to cut away around it and treat it like a hole, as follows.

Any perforation to the hull must be cut back until the edges are firm all around. Cracks may radiate out from even a small perforation, if impact has been severe, and the hole should be enlarged to the ends of such cracks. Once a firm edge has been attained, chamfer it on both sides so that the repair will key in successfully.

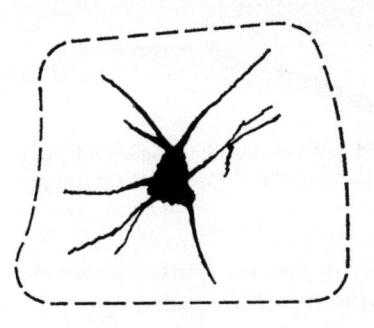

Cut away damage well outside extension of cracking. Shape of patch is immaterial.

Abrade the hull on both interior and exterior of the hole for an inch or so—just as for a bond.

You now have to blank off the hole with what is in effect a piece of a mould aligned with the area of hull surrounding the hole. If you are working on an area which is flat, or gently curved in one dimension, a piece of hardboard can be used—treated for release, of course. Or a piece of thin ply covered with polythene sheeting. Anything, in fact, which can be stuck with sticky tape or otherwise held in place firmly enough to let you laminate on top of it. If the repair is to a compound curve, a flat sheet cannot be bent to fit and you may have to conjure up a mould of plaster or Plasticene.

The mould patch should be affixed to the outside surface for repair so that you can use a coat of gel. This should be kept minimally thin so that the strength of repair lying in the lamination is unaffected. Laminate on top of the gel, making sure that the first layer of cloth is poked well down into the 'V' between patch and chamfer. Continue laminating normally, overlapping outward at the edges. Make the final lamination to stand a layer proud of its surrounds.

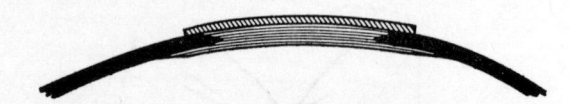

A patched hole. Note chamfering of edges and need to key repair material into outside notch.

When the repair has set hard you can detach the patch and tidy up the repair as needed.

MAJOR REPAIRS

I would define major damage as that which seriously affects the strength of the hull, or is of such a size or shape that an amateur could not reasonably expect to make a satisfactory mould for repair.

In such cases, which will probably be the subject of an insurance claim, the only sensible thing to do is to get a specialist firm to undertake the repair. This may be specified by the insurance company, but if possible you should get the work done by the original moulders. They will have the drawings, specifications and probably the mould from which the hull came, and be able to repair to a virtually new state.

REPAIRS AFLOAT

You will probably only be able to tackle minor damage incurred when afloat, and this is nearly always complicated by the presence of water in abundance. Holes can be plugged with rags or timber tapers wrapped around with rags. If a penetration is above the waterline, or can be got there by heeling the boat, a temporary resinglass patch can be put on. Or you could cut out a piece of ply or hardboard to cover the hole and bond around its edges; hardboard is best skinned over with a layer of glass. Small holes can be plugged with putty.

Somehow you will have to dry the area of repair as best you can, and swabbing down with acetone or styrene will assist.

Any repair carried out under such conditions is to be regarded as a purely temporary expedient, and the first opportunity taken to beach your craft for permanent repair.

14

Raising the Wind

There are several ways in which money can be raised for marine purposes, and in these days of inflation and high interests rates it is essential to get the best terms available. Perhaps the safest and cheapest of these is by way of a marine mortgage, obtainable from a Merchant Bank. Finance Houses tend to deal in loans and hire purchase, not mortgages. You can, of course, get an overdraft or personal loan from your own bank.

MORTGAGES

A marine mortgage is normally only obtainable in respect of a boat registered as a British Ship under the Merchant Shipping Act, 1896. This automatically excludes a bare or incomplete hull and other means of borrowing have to be sought. Use of some of them does not preclude your mortgaging the boat when completed and registered. The following advice was kindly supplied by Commander David Johnson of Hill, Samuel & Co:

'Being a Merchant Bank we handle only marine mortgage business on actual rates of interest payable on the day-to-day balance outstanding.

'When it comes to the question of someone purchasing a bare hull, then we normally take up our usual references and if these are suitable we can confirm to the customer's own bank our willingness to arrange a marine mortgage on the completed craft, once she has been so completed and registered with the Department of Trade and Industry. This normally enables the owner to have a bridging loan from his own Joint Stock Bankers, which we fill in as soon as registration has been completed and the mortgage signed.

'Some of the Finance Houses as opposed to Merchant Banks do offer some form of loan or hire-purchase facility to cover this point, but the snag to this arrangement is that the Finance Houses become the owners of the craft while it is subject to the hire purchase contract, and thus difficulties can arise in getting the craft registered eventually because it is not the owner's property. They are then faced with having to settle the HP agreement ahead of the period entered into and this usually results in a fairly hefty severance fee for foreclosing on the hire purchase deal.

'It is all rather complicated, but we have found that our normal method is perfectly satisfactory.'

I had dealings with Commander Johnson some years ago and know that his advice can be taken as sound and helpful—he is very knowledgeable about boats as well as finance.

HIRE PURCHASE

Apart from the ownership problems mentioned, some Finance Houses consider that loans in respect of kit and partially completed boats come into a difficult market. This is because they cannot know in advance the degree of skill which the purchaser will bring to bear on completion, and this aspect is made more difficult where bare hulls are concerned. However, they take a very different view on the fitting out of boats which are substantially completed when purchased. It will be of benefit for you to enquire widely if considering hire purchase.

BANK LOANS AND OVERDRAFTS

The tendency nowadays is for Joint Stock Banks to advance loans on a fixed-term basis. Capital repayment ahead of time will not result in a *pro rata* reduction of interest charges, as these are calculated initially and included as an element of the advance. Against this is the fact that the rate of interest is consequently fixed at the prevailing rate at the time of loan and is not subject to fluctuation, nowadays inevitably upwards.

Interest on overdrafts is calculated periodically on the outstanding balance, and repayments of capital at any time have the effect of reducing interest proportionately. Rates of interest will fluctuate over the period. An overdraft can, perhaps, be extended in both time and amount which is not so with a fixed-term loan—except perhaps in compassionate circumstances.

TAX RELIEF

Tax relief on loan interest is something for the politicians to conjure with—now you see it, now you don't. You would be crazy not to take advantage of it when it is available, and relief on a sizeable loan can be a significant amount. Local tax offices will, if harassed sufficiently, provide information on the existing position. It is also explained for the erudite in those revolting leaflets that accompany each year's firm and rapacious demand that you declare all. They are, of course, incomprehensible to the man in the street.

ACKNOWLEDGMENT

I am grateful to the following firms for advice given and information supplied. They can be safely approached about marine financing.

Hill, Samuel & Co., West End Office, 19, St. James' Square, London SW1Y 4JQ

Bowmaker Ltd., Bowmaker House, Christchurch Road, Bournemouth BH1 3LG

National Acceptances Ltd., Ramillies Buildings, 1–9, Hills Place, Oxford Circus, London W1R 2N H

Lombard North Central Ltd., 24 Cornwall Road, Dorchester, Dorset

Paybonds Ltd., P.O. Box 4, Water Lane, Bradford BD1 2JL

APPENDIX I

Properties and Use of Materials

In compiling this index I gratefully acknowledge the advice of the Strand Glass Company Ltd. of Brentwood, whose products I have successfully used for a number of years. As materials cannot be guaranteed to be precisely the same at all times, information is approximate, but should serve as a fair working guide for the amateur builder.

1. *Gelation time—polyester resin*

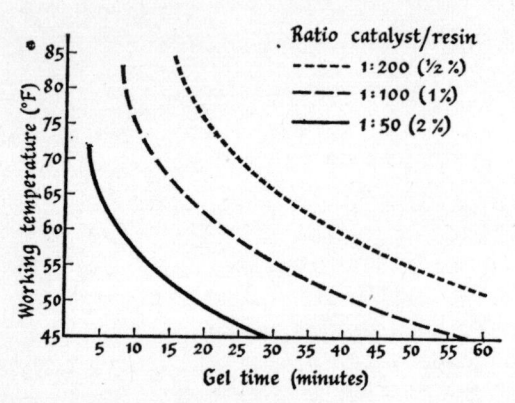

Ratio catalyst/resin
- ----- 1:200 (½%)
- – – – 1:100 (1%)
- ——— 1:50 (2%)

x-axis: Gel time (minutes); y-axis: Working temperature (°F)

In working temperatures of 80°F and over, catalyst must be used in small percentage, never more than ½ per cent.

2. *Gelation time—epoxy resin*

Catalyst must be added in invariable proportion as

PROPERTIES AND USE OF MATERIALS

recommended. Gel time is about 6hr at 45°F, 2hr at 65° and 1hr at 85°. Generally, epoxy resin should remain undisturbed for 24hr so that it sets firm.

3. *Catalyst for polyester—percentages*

Liquid (50 per cent MEKP): 1 per cent = 5cc or 1/6oz to 1lb of resin.

Paste: 1 per cent = 6 inches from tube or 1/3oz to 1lb of resin.

It will be obvious that paste is only half as strong as liquid. For small quantities of resin a paste catalyst can therefore be more accurately measured.

4. *Working temperatures*

It is quite possible to use resin mixes in cold weather, but then a much greater percentage of catalyst has to be added to permit gelation in a reasonable time. Irrespective of ambient temperature, a large proportion of catalyst will tend to cause the mix to heat up, and attention must be given to possible combustion.

Curing, as distinct from gelling, is a relatively lengthy period. At 60°F it takes ordinary resin about a week to cure but there are quick-curing resins available. At lower temperatures curing may be unduly prolonged with risk of a distorted set. This is not serious where bonds are concerned, but beware of mouldings going out of shape in cold weather.

5. *Thickness of laminates*

A fully wetted-out layer of woven rovings, tape or cloth will be the same thickness as the dry cloth.

CSM beds down on wetting and the laminate will, if well rolled, be thinner than the dry cloth. A wetted-out and rolled layer of $1\frac{1}{2}$oz CSM is about 1/16 inch thick

FITTING OUT A MOULDED HULL

and other mats are proportional, i.e. 6oz C S M laminate is some ¼ inch thick.

6. *To bond wooden fitments*

Wood should be dry, clean and roughed up. Adhesion is excellent if the surface to be bonded is then given a brushing with an accelerator such as cobalt naphthenate. This should dry off thoroughly before the resin is applied.

Strength members should be bonded with an aggregate of lamination equal in thickness to the hull at the line of join, say 10oz for a 20-foot boat. If bonded on one side only the laminate should be that thick, but if on both there should be 5oz on either side.

Non-critical joins can be made with two layers of 1½oz C S M.

For rigid joins use C S M next to the wood and hull, followed by a layer of 5oz woven cloth. For flexibility use woven rovings, 8oz, on top of the mat.

7. *Tanks*

A five-gallon tank can be moulded with a layer of 1½oz C S M, one of 8oz woven rovings and a final one of 1½oz C S M. The same lamination will serve for tanks up to 25 gallons, provided that thin aluminium strips are laminated in about eighteen inches apart. These can conveniently be carried around edges.

8. *Strength of resinglass, when fully cured*

	Tensile strength lb per sq in	Compression strength lb per sq in
Polyester:	6,000 – 13,000	13,000 – 36,500
Epoxy:	6,000 – 13,000	15,000 – 25,000

PROPERTIES AND USE OF MATERIALS

9. *Types of resin available (Strand Glass)*

Gel
Lay-up
Clear casting
Translucent
Flexible
Fire retardant (two grades)
Heat and chemically resistant
Fire resistant for coating existing surfaces
Epoxy

Mixtures are obtainable if you wish to combine properties, e.g. a flexible, translucent, fire retardant resin can be achieved.

10. *Costs*

Resin gets cheaper as you buy larger quantities, the difference being in problems of breaking it down into small quantities. For example, a lay-up resin costing 20p a lb when bought in a 3lb tin costs 13p a lb for a cwt and 10p a lb for two 496lb drums.

Glass is similarly scaled down, but not to the degree that will be of much help to a casual user of small quantities.

It took me about 1cwt of lay-up resin to fit out my 25-foot hull fully and it is well worth buying such an amount in bulk if you can store it safely. At present prices (1973) two 56lb drums cost £14.56, whereas the same quantity in 10lb tins costs £20.10 and in 3lb tins, £22.40.

11. *Resin required for full wet-out*

	type of glass	weight of resin
CSM:	1oz (per sq ft)	1½lb per sq yd
	1½oz ,, ,, ,,	2lb ,, ,, ,,
	2oz ,, ,, ,,	3lb ,, ,, ,,

FITTING OUT A MOULDED HULL

Fabrics: 2oz (per sq yd) $\frac{3}{4}$lb ,, ,, ,,
           5oz   ,, ,, ,,   1lb ,, ,, ,,
           8oz   ,, ,, ,, $1\frac{1}{2}$lb ,, ,, ,,
          24oz ,, ,, ,, $4\frac{1}{2}$lb ,, ,, ,,

12. *Miscellaneous requirements*

Gel-coat: $1\frac{1}{4}$lb of resin per sq yd of mould surface.
P V A release agent: 2oz per sq yd of mould surface.
Colouring pigment: 1lb per 14lb of resin.
Pre-gel paste: 1lb per 12lb of resin.

APPENDIX II

Availability of Hulls

As stated in the text, there are innumerable firms producing mouldings, ranging from good to indifferent, which advertise in such periodicals as *Practical Boat Owner* or whatever you read. Those which are listed below are representative of the better-known suppliers from whose products most readers might expect to find a boat to suit their needs.

SAILING BOATS

FIRM	PRODUCTS
Ardleigh Laminated Plastics Ltd., Wheaton Road, Industrial Estate East, Witham, Essex.	Large range of fin, twin and c/b hulls and superstructures for D I Y completion. Bonded together on request.
B E B S (Marine) Ltd., Quay Works, Burnham-on-Crouch, Essex.	Large range of hulls and superstructures in almost any stage for D I Y completion. Also supply steel hulls.
Galion Yachts Ltd., Bridge Road, Bursledon, Hants.	22 – 27ft fin keel boats. Hulls and kits.

FITTING OUT A MOULDED HULL

Halmatic Ltd., Brookside Road, Havant, Hants.

A large range of hulls and superstructures eminently suitable for DIY completion.

Hunter Boats Ltd., Sutton Wharf, Sutton Road, Rochford, Essex SS4 1LZ.

16 – 24 ft fin and twin. Hulls and kits.

Island Plastics Ltd., Edward Street, Ryde, I.O.W.

Large range of hulls, superstructures and modules.

Macwester Marine Co. Ltd., River Road, Littlehampton, Sussex BN17 5D B.

22 – 32ft fin and twin. From bare hull to part-complete.

Marine Construction (UK) Ltd., Willments Shipyard, Hazel Road, Woolston, Southampton SO2 7G B.

22 – 30ft fin and twin. Hulls and superstructures supplied to various stages of completion.

Newbridge Boats Ltd., Church St., Bridport, Dorset.

22 – 23ft fin and twin. Hulls to various stages of completion. Kits of parts.

Offshore Yachts, Mill Road, Royston, Herts.

Large range of fin and twin. Hulls and kits.

J. C. Rogers, Boatbuilder, Waterloo Road, Lymington, Hants.

26 – 32ft fin. From bare to complete as required.

Russell Marine Ltd., Calvia Works, Prince

Large range of 'family' fin and twin. Hulls with

AVAILABILITY OF HULLS

Avenue, Southend-on-Sea, Essex.	kits supplied in stages if required.
Rydgeway Marine Ltd., Church Road, Kessingland, Nr. Lowestoft.	19 – 22ft fin, twin and c/b. Hulls and kits. Modules.
Seamaster Ltd., 20 Ongar Road, Great Dunmow, Essex CM6 1EU.	19 – 23ft fin or c/b. Semi-fitted hulls supplied.
Tyler Boat Co. Ltd., Sovereign Way, Tonbridge, Kent.	Large range of 22 – 40+ ft cruisers. Normally supplied in fairly complete form with keel, bulkheads, rudder, engine bearers, etc. fitted.
Westerly Marine Construction Ltd., Aysgarth Road, Waterlooville, Portsmouth PO7 7UF.	21 – 32½ft fin and twin. Hulls and kits.

POWER CRAFT

Ardleigh Laminated Plastics. As above.	Large range of cruising hulls and superstructures. Engines can be supplied fitted if required.
Aquaboats Ltd., 261 – 263 Lymington Road, Highcliffe-on-Sea, Christchurch, Hants.	Fair range of cruisers in various stages of completion.
Essex Yacht Brokers Ltd., Wallasea Island, Nr. Rochford, Essex.	Cruiser hulls and kits in any stage of completion.

FITTING OUT A MOULDED HULL

Halmatic Ltd. As above.	Extensive range of cruisers in hull and module form.
Island Plastics. As above.	Good range of hulls, superstructures and modules for D I Y completion.
Macwester Marine. As above.	26ft fishing boats in various stages of completion.
Monachorum Mfring. Co. Ltd., Wixenforn Farm, Coleshill Down, Plymouth, Devon.	16 – 18ft fishing boats in hull and kit form.
Newbridge Boats. As above.	32ft cruisers in various stages.
Seamaster Ltd. As above.	23 – 30ft cruisers in various stages of completion.
Smallcraft (Portsmouth) Ltd., 348 Commercial Road, Portsmouth, Hants.	Small cruisers in various stages of completion.
Teal Boats Ltd., Langton St., Heywood, Lancs.	18 – 22ft cruisers bare or part complete.

APPENDIX III

List of Recommended Reading

Handyman Afloat and Ashore. K. Bramham. (Coles).
Yacht Construction. K. H. C. Jurd. (Coles).
Boat Building from Kits. K. Mason. (Bosun Books).
Designers Notebook. Ian Nicholson. (Coles).
Fibreglass Boats. Hugo du Plessis. (Coles).
G R P Foam Sandwich Construction. John Sears. (Coles).
Complete Amateur Boat Building. M. Verney. (J. Murray).

Index

Abrading, 25, 35-6, 154
Accelerator, 15
Acetone, 26, 37, 155

Backing pads, 30, 33, 125, 134
Barrier cream, 16, 37
Bonding, 25-8, 41, 118, 162
Bulkheads, 20, 45, 47, 72-3, 75, 113-18

Catalyst, 15-19, 45, 161
Chainplates, 12, 102-3
Chopped strand mat (C S M), 13, 23, 66
Cleanliness, 36, 39
Cloth, woven, 13, 31
Cloth, glass-, types of, 13-14
Cloth, glass-, use of, 25-9, 92, 161-2
Coamings, 100-101, 149
Cockpit drains, 92-3, 145
Condensation, 129-131

Costs, 46-7, 156-9
Curing time, 18, 25

Decoration, 39-41, 129
Design, 13, 44, 52
Distortion, 18, 25, 47
Drainage, 76

Engine, 50-1, 134-144
Engine bearers, 136-8
Engine cooling systems, 140-1
Epoxy glue, 32
Epoxy resin, 16, 32, 146
Exhaust pipes, 142-3
Explosion, 15-16

Fire risk, 41, 142-3
Fire precautions, 41, 120, 143
Fittings, 21, 42, 48, 54, 76, 93, 148
Framing, 119-20

171

Galley, 120
Gelation, 15, 17, 160
Gel-coat, 19, 21, 43, 79-82
Glue, 32, 40-1
Grease, 25, 27
Green stage, 17, 47
Ground tackle, 125-7

Handling characteristics, 48-9
Hatch cover, 77-80, 93-6
Hinges, 93-4, 121
Hull datum lines, 70-1, 73
Hull, resinglass, 44-6

Keels, 90-2

Laminate, 12, 17, 20
Lamination, 12, 25, 29, 43
Laying up, 12, 20, 36, 83
Leaks, 96-7
Level, 54, 70, 73
Level, spirit-, 71, 73, 114
Lights, navigation, 107, 146
Lights, window and other, 131-3
Lockers, 49, 121-3

Maintenance, 149-152
Mast step, 104
Mould, male, 21

Mould, female, 21, 23, 43-4, 77-81
Moulding attachments, 21-3
Moulding a hatch cover, 77, 79-81, 93-5
Moulding certificate, 53
Moulding, distortion of, 18, 25, 47
Moulding, hull, 25, 43, 45-6. 88
Moulding release, 23, 81, 84
Moulding tanks, 21
Moulding toerails, 99-100, 106, 149

Paint, 39-40, 152
Performance, 49
Pigment, 19, 38, 45, 152
Piping, 111-12
Plug, 43-4
Polyester resin mix, 16
Polythene, use of, 21, 36, 107
Power, 50, 65
Price, 52, 56
Protection, 150-2
Pulpits, 105-6
Pumps, 145

Release, moulding, 23, 81, 84
Release agents, 21-2, 81-2
Repair, 11, 152-5
Resin, epoxy, 15, 28, 96
Resin, polyester, 15, 19, 28, 45, 89

INDEX

Resin, thinned, 27
Resinglass, 11-14
Resinglass, use of, 20, 80, 82, 89, 91-2
Resinglass, properties of, 18-19, 160-4
Rigging, 48, 54
Rubbing strakes, 45, 47, 86-90
Rudder, 108-9

Sandwich construction, 33-4
Seaworthiness, 48
Seacocks, 125, 145
Setting times, 16, 24, 27, 28
Shelf life, 15, 55
Slings, 67-9
Spiling, 113-18
Stanchions, 12, 105-6
Stowage, 49, 66, 125
Stringers, 12, 20, 118
Stresses, 29-30, 33, 108
Styrene, 26-7, 38

Symmetry, 73-4, 76

Tanks, water, 11-12, 21, 84-5, 128
Tanks, fuel, 22, 129, 136
Templates, 66, 113-18
Thixotropic agent, 20, 26
Toerails, 13, 100
Tools, 34-5, 37, 42, 57-9, 66, 70
Tracking, 96, 105

Value, 53
Ventilation, 98-9, 127-8

Waterline, 70-2, 73
Wetting-out, 13-14, 20, 24, 45, 78, 82
Winches, 13, 101-2
Windows, 131-3
Wiring, 111-12
Working schedule, 61-4, 86